T0301847

Understanding Jitter and Phase Noise

A Circuits and Systems Perspective

Gain an intuitive understanding of jitter and phase noise with this authoritative guide. Leading researchers provide expert insights on a wide range of topics, from general theory and the effects of jitter on circuits and systems, to key statistical properties and numerical techniques. Using the tools provided in this book, you will learn how and when jitter and phase noise occur, their relationship with one another, how they can degrade circuit performance, and how to mitigate their effects – all in the context of the most recent research in the field. Examine the impact of jitter in key application areas, including digital circuits and systems, data converters, wirelines, and wireless systems, and learn how to simulate it using the accompanying Matlab code. Supported by additional examples and exercises online, this is a one-stop guide for graduate students and practicing engineers interested in improving the performance of modern electronic circuits and systems.

Nicola Da Dalt is Analog Engineering Manager for High-Speed Serial Interfaces at Intel Corporation, having previously worked at Telecom Italia and Infineon Technologies.

Ali Sheikholeslami is Professor in the Department of Electrical and Computer Engineering at the University of Toronto.

As we continue to push operating speeds in electronic systems, timing jitter has emerged as an increasingly important showstopper across a wide range of applications. Consequently, pushing the envelope requires a thorough understanding of jitter from its mathematical description, to its manifestation in circuits and its impact on systems. This book delivers the most comprehensive treatment of this subject to date and provides valuable content to jitter-plagued engineers at all levels of experience.

Boris Murmann, *Stanford University*

All components generate noise. They give rise to thermal and 1/f noise. All amplifiers and filters have Signal-to-Noise ratio as one of their most important specifications. In oscillators however, noise gives rise to jitter and phase noise. This is why this book is so important. It provides unique insight in the origins and the analysis of these specifications. Many applications are highlighted in the field of data converters, wireless and wireline systems, and a number of digital applications. Examples are the jitter in a CMOS inverter, in a LC oscillator, in a ring oscillator, etc. As a result this book is a necessity for all designers who have to know about noise and its performance limitations.

Willy Sansen, *KU Leuven*

Phase noise is the primary source of performance deterioration in all wireless/wireline communication systems – and yet, dedicated books have been conspicuously absent to date. We are therefore very fortunate that two real experts – Dr. Da Dalt and Professor Sheikholeslami – have finally decided to fill this gap, presenting us with what will become standard reading for anyone desirous to understand the peculiar and often elusive nature of phase noise.

Professor Pietro Andreani, *Lund University*

The rigorous mathematical description of jitter, its link to phase noise as well as its practical impact on different classes of circuits (e.g. digital, wireline, wireless, data converters) are all known as difficult and sometimes obscure topics even for experienced designers. This is the only book that I know which covers all of these subjects, providing at the same time both the intuitive understanding, the Matlab codes are particularly useful from this standpoint, and the appropriate mathematical rigour. The authors, that are two leading experts in the field, have also done a significant effort also in discussing the key findings available in both classical and more recent open literature, not just presenting their own work. I highly recommend this book.

Carlo Samori, *Politecnico di Milano*

This excellent reference provides a wealth of material to satisfy both engineers new to clocking and seasoned veterans that are experts in jitter and phase noise. The authors address all the important aspects of these critical topics and provide great insights for readers.

Samuel M Palermo, *Texas A&M University*

Understanding Jitter and Phase Noise

A Circuits and Systems Perspective

NICOLA DA DALT
Intel Corporation

ALI SHEIKHOLESLAMI
University of Toronto

CAMBRIDGE
UNIVERSITY PRESS

University Printing House, Cambridge CB2 8BS, United Kingdom

One Liberty Plaza, 20th Floor, New York, NY 10006, USA

477 Williamstown Road, Port Melbourne, VIC 3207, Australia

314-321, 3rd Floor, Plot 3, Splendor Forum, Jasola District Centre, New Delhi - 110025, India

79 Anson Road, #06-04/06, Singapore 079906

Cambridge University Press is part of the University of Cambridge.

It furthers the University's mission by disseminating knowledge in the pursuit of
education, learning and research at the highest international levels of excellence.

www.cambridge.org
Information on this title: www.cambridge.org/9781107188570
DOI: 10.1017/9781316981238

First published 2018

A catalogue record for this publication is available from the British Library

Library of Congress Cataloging in Publication data
Names: Da Dalt, Nicola, 1969– author. | Sheikholeslami, Ali, 1966– author.
Title: Understanding jitter and phase noise : a circuits and systems perspective /
    Nicola Da Dalt (Altera Corporation), Ali Sheikholeslami (University of Toronto).
Description: Cambridge, United Kingdom ; New York, NY : Cambridge
    University Press, 2018.
Identifiers: LCCN 2017031090| ISBN 9781107188570 (hardback) |
    ISBN 1107188571 (hardback)
Subjects: LCSH: Electric circuits–Noise. | Electric noise. | Oscillators, Electric–Noise.
Classification: LCC TK454 .D3 2018 | DDC 621.382/24–dc23
LC record available at https://lccn.loc.gov/2017031090

ISBN 978-1-107-18857-0 Hardback

To our parents,
Giuliana and Guido
Fatemeh and Hadi

Contents

Preface

This book provides a rigorous yet intuitive explanation of jitter and phase noise as they appear in electrical circuits and systems. The book is intended for graduate students and practicing engineers who wish to deepen their understanding of jitter and phase noise, and their properties, and wish to learn methods of simulating, monitoring, and mitigating jitter. It assumes basic knowledge of probability, random variables, and random processes, as taught typically at the third- or fourth-year undergraduate level, or at the graduate level, in electrical and computer engineering.

The book is organized as follows: Chapter 1 provides a qualitative overview of the book and its contents. Chapter 2 covers the basics of jitter, including formal definitions of various types of jitter and the key statistical concepts, starting from jitter mean and the standard deviation up to random and deterministic jitter. Phase noise will be first introduced in Chapter 3, and its relation to jitter and to the voltage spectrum of the clock signal will be extensively investigated. In particular, how to derive from phase noise the values of the several jitter types introduced previously will be explained. Chapter 4 is dedicated to the effects of jitter and phase noise in basic circuits and in basic building blocks such as oscillators, frequency dividers, and multipliers. Chapters 5 to 8 discuss the effects of jitter and phase noise in various circuit applications. Chapter 5 is dedicated to the effects of jitter on digital circuits, Chapter 6 to data converters, Chapter 7 to wireline, and Chapter 8 to wireless systems. More advanced topics on jitter are covered in Chapter 9, followed by numerical methods for jitter in Chapter 10. This chapter also explains how to generate jitter and phase noise, with various characteristics, for simulation purposes. The corresponding Matlab code for producing jitter is included in Appendix B.

As mentioned earlier, this book assumes the reader has a basic knowledge of random variables and random processes. However, to refresh the reader's memory of the definitions of some key terms, Appendix A simply lists these key terms along with their basic definitions.

Guidance for the Reader

The book does not require the reader to adhere strictly to the order in which the chapters appear, nor to read all of them. Its structure and the content of each chapter allow different paths to be followed, depending on the particular interests or learning objective of the reader. The graph below summarizes the possible paths, with the solid boxes indicating strongly recommended chapters and the dashed boxes the suggested additional readings.

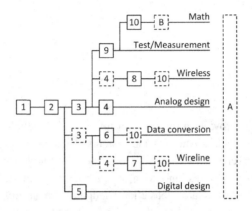

While Chapters 1 and 2 form the fundamentals, and thus should be read before any other chapter, the remaining chapters are relatively independent from each other. Chapter 3 introduces the concept of phase noise and its relation to jitter. Even though this chapter constitutes, together with Chapter 8, a required path to the reader active in the wireless field, its contents are relevant to a number of other application fields, among them wireline and jitter testing. For this reason the authors suggest it should be included independently of the particular focus. Chapter 4 is an important reading for analog IC designers, while Chapter 5 addresses specifically the needs of custom digital designers. The latter chapter does not require knowledge of phase noise; thus Chapter 3 could be omitted. Both Chapters 4 and 5 can be skipped by readers interested exclusively in the system or mathematical aspects of jitter and phase noise. Chapters 6 and 7 can be read directly after the first three chapters by readers interested in data converters or wireline communication systems respectively. For the reader whose interest lies in the mathematical treatment of jitter and phase noise, the first three chapters plus Chapter 9 will provide a complete path. Finally, Chapter 10 and Appendix B are suggested reading for students or engineers who want to analyze the effect of jitter and phase noise on systems of any nature by means of transient simulation. The book uses a number of terms from probability and random processes. For ease of reference, we have included these key terms and their brief definitions in Appendix A.

Acknowledgments

In writing this book we were fortunate to receive support from numerous professors, colleagues, and friends from both academia and industry. The quality of the book was greatly improved by their innumerable suggestions, questions, and discussions. For this, we sincerely thank and are deeply indebted to Pietro Andreani (Lund University), Yunzhi (Rocky) Dong (Analog Devices), Salvatore Levantino (Politecnico di Milano), Antonio Liscidini (University of Toronto), Gabriele Manganaro (Analog Devices), Boris Murmann (Stanford University), Maurits Ortmanns (Ulm University), Samuel Palermo (Texas A&M University), Behzad Razavi (University of California, Los Angeles), Carlo Samori (Politecnico di Milano), Willy Sansen (KU Leuven), Richard Schreier (Analog Devices), Hossein Shakiba (Huawei), and John Stonick (Synopsys). To them the merit of having made this a better book. The mistakes are all ours.

We wish to thank also Elad Alon (University of California, Berkeley), Seong Hwan Cho (Korea Advanced Institute of Science and Technology), Ali Hajimiri (California Institute of Technology), Hirotaka Tamura (Fujitsu Labs of Japan), and Tadahiro Kuroda (Keio University), who reviewed the book proposal and made valuable suggestions.

A number of graduate students from the University of Toronto provided considerable help in reading and commenting on various sections of this book. We would particularly like to express our appreciation to Keivan Dabiri, Joshua Liang, Zeynep Lulec, Mehrdad Malekmohammadi, Wahid Rahman, Luke Wang, and Alif Zaman.

We are indebted to Sara Scharf from University of Toronto, who kindly agreed to review and edit the entire original manuscript.

A special thank-you to our families, who encouraged and supported us during the long period of thinking and writing. Thank you for your constant love, for taking care of all the rest, and for making sure that, outside of this book, life still went on as usual – and for enduring the nights, weekends, and vacations subtracted from our time together.

1 Introduction to Jitter

Imagine a lunar eclipse is to occur later tonight at around 1:57 am when you are fast asleep. You decide to set your camera to capture the event, but being uncertain of the actual time, you would set the camera to start a few minutes before the nominal time and run it for a few minutes after 1:57 am. In this process, you have included a margin around the nominal time so as to minimize the risk of not capturing the event on your camera.

Imagine you have an important meeting at 8:00 am sharp, but you are uncertain about your watch being a bit too fast or too slow. Determined not to be late, you decide to arrive by 7:55 am according to your watch, just in case your watch is a bit slow. This is indeed a practical way to deal with uncertainty in time.

Imagine you have a ticket for the bullet train in Tokyo that leaves the train station at 1:12 pm sharp. You are at the station and on the right platform, and you take the train that leaves the station at 1:12 pm sharp. But a few minutes later, you notice that your watch is two minutes fast compared to the time being displayed on the train, and you realize that you are on the 1:10 pm train and you are moving towards a different destination.

Missing an event, being late or early, and getting on the wrong train are all consequences of timing uncertainties in our daily lives. In digital circuits, we deal with very similar situations when we try to time events by a clock that has its own timing uncertainty, called *jitter*, and in doing so we may miss an event, such as not capturing critical data, or cause bit errors, e.g., capturing the wrong data. Our goal, however, is to prevent such errors from occurring, or to minimize their probabilities. We do this first by carefully studying the nature of these uncertainties and modeling their characteristics.

This chapter provides a few concrete examples from electronic circuits and systems where timing uncertainties have a profound impact on the accuracy of their operation. Our goal here is to qualitatively introduce the concept of jitter and intuitively explain how it should be characterized and dealt with. A formal definition of jitter, its types, and its full characterization will be presented in Chapter 2.

1.1 What Is Clock Jitter?

This section provides an intuitive explanation of two fundamental jitter concepts: period jitter and absolute jitter.

1.1.1   Period Jitter

Most microprocessors (μP) and digital signal processing (DSP) units work with clock signals to time the execution of instructions. A basic instruction such as a shift by one bit may take a single clock cycle, whereas addition and subtraction instructions may take three to four clock cycles, and more complex instructions, such as multiplication, may take tens of clock cycles to complete. In all these cases, the underlying assumption is that the clock is accurate; that is, the clock cycles are all identical in duration. In reality, however, the cycles of a clock are only identical with a finite (not infinite) accuracy. This is because the clock signals are generated using physical devices that inevitably have some uncertainty or randomness associated with them. For example, a clock may be generated by three identical CMOS inverters in a loop, as shown in Figure 1.1. If we assume t_{pd} is the propagation delay of each inverter in response to a transition at the input, then it takes $3t_{pd}$ for a state "0" at node v_{out} to transition to state "1" and another $3t_{pd}$ to transition back to "0". In total, it takes $6t_{pd}$ for the clock to complete one cycle, and therefore we can write $T_{nom} = 6t_{pd}$ where T_{nom} represents a nominal clock cycle. However, strictly speaking, t_{pd} is not constant, but rather a random variable that results in a different propagation delay every time the inverter is used. This randomness is inherent in any physical device that deals with the motion of electrons at temperatures above zero degrees Kelvin.

If we accept that t_{pd} is a random variable, then the clock period, which is the sum of six random variables, is also a random variable that deviates from its nominal (or expected) value. The good news is that these random variables may have tight distributions and, as such, may not adversely affect the circuit operations, especially at low clock frequencies where the period is much larger than the deviations. However, as we increase the clock frequency, the same absolute deviation in period may compromise circuit operation.

A typical clock frequency of 4GHz, for example, corresponds to a nominal period of 250ps, but this period may have a normal (Gaussian) distribution with a standard

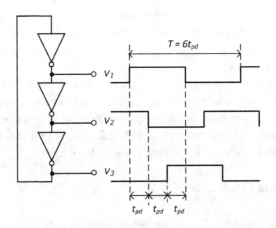

Figure 1.1 A simple ring oscillator and its voltage waveforms.

deviation of $\sigma = 1\mathrm{ps}$. This implies that a clock cycle is close to 250ps but may deviate from this value with a probability that is a decreasing function of the deviation magnitude. For example, as we will see in Chapter 9 (Table 9.1), the probability of the period being smaller than 243ps is about 10^{-12}. What does this mean for designs using this clock? A straightforward answer is that we must ensure that any task that is supposed to complete in one clock cycle must do so assuming the worst-case clock cycle, that is, 243ps (not the nominal 250ps). This will ensure that, with a high probability $(1 - 10^{-12})$, all the tasks are completed within one clock cycle even when one clock cycle is 7ps shorter than the nominal cycle. If we were to reduce the probability of failure (i.e., the probability of failing to meet the timing), we could simply design for a 242ps minimum cycle time while working with a nominal cycle time of 250ps. This reduces the probability of failure to 6.22×10^{-16}, which corresponds to one error in every 4.6 days (assuming 4GHz operation).

The concept we just described, i.e., the deviation of a clock cycle from its nominal value, is referred to as *clock period jitter*, or *period jitter* for short. A concept closely related to period jitter is *N-period jitter*, which is the deviation of a time interval consisting of N consecutive periods of the clock from the time interval of N nominal periods. We will formally define period jitter, N-period jitter, and other types of jitter for a clock signal in Chapter 2. Let us now provide an intuitive explanation for *absolute jitter*.

1.1.2   Absolute Jitter

Period jitter is only one method to characterize the timing uncertainties in a clock signal. This method is particularly useful in digital circuits where we are concerned with the time period we need to complete a task. In other applications, such as in clock and data recovery, where the clock edge is used to sample (capture) data, we are concerned with the instants of time, not with time durations! This is because sampling occurs at an instant of time, not over a period of time. In this case, it is the absolute time of the clock edge (at which sampling occurs) that matters, not the interval between the edges (i.e., the clock period). In these applications, it is important for the clock's edges to be precise; i.e., for them not to deviate from their ideal locations by certain amount. Let us elaborate on this further.

In an ideal clock, if we assume the first rising edge occurs at time $t = 0$, then the subsequent rising edges will occur at exactly $t = kT$, where k is a positive integer and T is the period of the ideal clock. In a non-ideal clock, the rising edges, t_k, will deviate from their ideal values kT. We refer to this deviation as the clock's *absolute jitter*, with the word *absolute* signifying the deviation in instants of time as opposed to deviations in intervals of time, as in period jitter.

The reader notes that absolute jitter and period jitter are closely connected concepts. Indeed, we will see in Chapter 2 how we can obtain one from the other.

Figure 1.2 shows an example of a jittery clock signal (CK) along with an ideal clock signal (CK_{ideal}). The absolute jitter can be abstracted from the clock signal and shown separately as a function of time. From this example, absolute jitter appears to be a

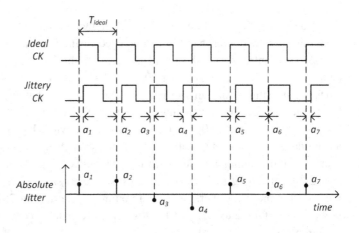

Figure 1.2 Ideal clock versus jittery clock waveforms.

discrete-time random signal. We will see later in Chapter 2 that, similar to any random signal, absolute jitter has a well-defined spectrum and, as such, may contain a continuous range of frequency components.

1.1.3   Intentional Jitter

Jitter is not always an unwanted property of a clock signal. In some applications, such as in spread-spectrum clocking, we intentionally add carefully controlled jitter to a clean clock so as to shape its power spectral density.

It is well known that a clean clock, by virtue of being a periodic signal, has energy peaks at its fundamental frequency and its second- and higher-order harmonics. At clock frequencies in excess of 100MHz, some of the harmonics may have a wavelength that is of the same order of magnitude as the length of the wire carrying the clock signal between chips on the same board or between boards. If left unattenuated, the signal energy of these harmonics may radiate and cause electromagnetic interference (EMI) with neighboring electronic components (mostly the components on the same PCB). For this reason, it is desirable to limit the amount of radiation a clock signal can produce. This radiation level is directly related to the peaks in the clock signal power spectrum.

Spread-Spectrum Clocking is a technique that spreads the peak energy over a larger frequency band, thereby reducing the peak values and their associated EMI. This is accomplished by adding a controlled amount of jitter to the clock. The jitter profile is typically in the form of a frequency offset that goes up and down linearly with time. Figure 1.3 shows an example of a frequency offset which increases linearly over a certain period of time (kT) and then decreases over the same period (kT). This increase in clock frequency (assuming $\Delta f \ll f_0$) is equivalent to a decrease in the period by a factor of $2\Delta f / f_0$.

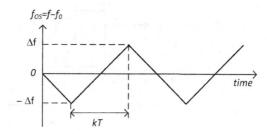

Figure 1.3 Frequency offset (f_{OS}) as a function of time.

Another example of using intentional jitter is in characterizing clock and data recovery circuits for their robustness to jitter. In this application, we intentionally add jitter to the incoming data and observe the system's robustness (in terms of its bit error rate) as a function of jitter frequency. We discuss this example in further detail in Chapter 7.

1.2   What Is Data Jitter?

Consider a D flip-flop with data input D, clock input CK, and output Q, where D is captured and transferred to Q at the rising edge of the clock. As a result, Q is synchronized with the clock, and therefore the clock jitter directly affects the timing of the data at the output. In clock and data recovery applications, it is common to refer to the duration of one data bit as a unit interval or UI. An ideal clock results in a data stream that has a fixed UI, where each UI is exactly equal to the clock period. We refer to this data stream as *ideal* or *clean* (i.e., the data without jitter). A non-ideal clock results in a jittery data stream where each UI is slightly different from the nominal UI. This situation is shown in Figure 1.4, where jittery data is compared against an ideal data stream. Note that, due to the random nature of the data sequence, data transitions do not occur at every UI edge. As such, jitter is not observable from data when there is no transition. However, by overlaying several UIs of data on top of each other, in what is known as an eye diagram, we can observe the overall characteristics of data jitter.

1.2.1   Eye Diagram

In clock and data recovery applications, the clock samples a jittery data waveform at instants that are one clock period (T) apart. These instants usually correspond to the rising edges of a clean clock, i.e., a clock with little or no jitter. Since the act of sampling repeats itself every T, we are interested in seeing how the data waveform looks if we chop it into segments of length T and overlay all the segments on top of each other. The resulting composite waveform, which resembles an eye, is known as an eye diagram.

Figure 1.5 shows the process by which we generate an eye diagram. This is essentially the data waveform as a function of time modulo $1T$. The eye diagram clearly shows the

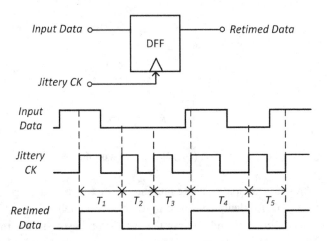

Figure 1.4 Input data is retimed with a jittery clock.

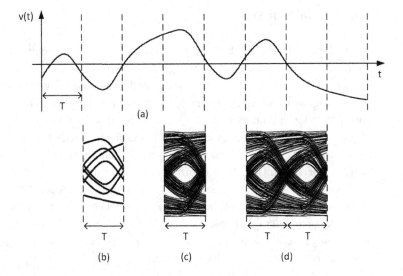

Figure 1.5 (a) A sketch of a voltage waveform as a function of time, (b) an eye diagram with eight traces, (c) an eye diagram with 500 traces, (d) an eye diagram showing two eyes (corresponding to two periods).

time instants at which the data "0" and "1" are farthest apart. This is the instant at which the eye height (also known as vertical eye opening) is at a maximum, and this is the instant at which we strive to sample the data by the rising edge of the clock.

The eye diagram also reveals how jitter can accumulate over time so as to close the eye in the horizontal direction (see Figure 1.6). Accordingly, the horizontal eye opening is an indication of the timing margin left for error-free operation.

When the noise, interference, attenuation, and jitter are excessive in a design, the eye may be completely closed: there is then no instant at which the data "1" and "0" can be

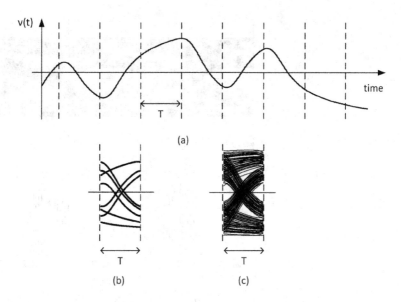

Figure 1.6 (a) A sketch of a voltage waveform as a function of time, (b) an eye diagram centered around the zero crossings (eight traces), (c) the same eye diagram using 500 traces.

consistently distinguished. These situations require equalization and jitter cleaning so as to open an otherwise closed eye for data detection.

1.2.2 Random Versus Deterministic Jitter

Jitter, in general, refers to any timing deviations from an ideal clock. These deviations may be random or deterministic depending on the underlying sources producing them. We have already provided several examples where random jitter is produced. Let us now consider an example that produces deterministic jitter. Consider a binary random sequence (produced using an ideal clock) that travels the length of a wire on a PCB. The wire acts as a low-pass filter and, as such, slows the data transitions, creating jitter. The jitter produced in this way is considered deterministic because one can predict the exact jitter if an accurate model of the wire is available. Note that this is in contrast with the random jitter produced by random movement of electrons, as we cannot predict the electron movement.

An easy way to characterize jitter is to look at its histogram or its probability density function (PDF). As shown in Figure 1.7, deterministic jitter is bounded (having a histogram with limited range) whereas random jitter is unbounded (having a Gaussian-like PDF). In most cases, several jitter sources are at work at the same time, producing composite jitter that includes both deterministic and random jitter. We will discuss this in depth in Chapters 2 and 7 and describe ways to decompose jitter into its components. Independent of the type of jitter, however, one can always characterize jitter, which is a discrete-time sequence, by its peak-to-peak or its root mean square (RMS) values.

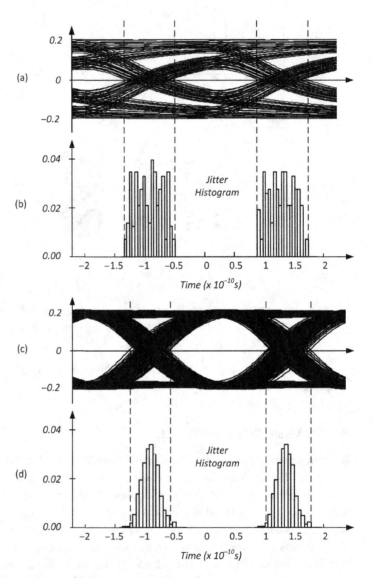

Figure 1.7 Eye diagrams and their corresponding jitter histograms (a) eye diagram with deterministic jitter only, (b) corresponding jitter histogram, (c) eye diagram with random jitter only, (d) corresponding jitter histogram.

1.3   Jitter in Measuring Time

We often measure time by observing the completion of a physical phenomenon or event. For example, in an hourglass (also known as a sandglass), shown in Figure 1.8, the event is the falling of sand from the top to the bottom compartment. The completion of this event marks one unit of time, which may be from a few minutes to a few hours depending on the hourglass design. Note that an hourglass has limited

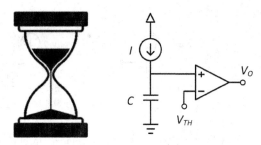

Figure 1.8 An hourglass behaves similarly to a current source charging a capacitor.

accuracy and limited resolution. The accuracy relates to the variation in time measurement as we repeat the same experiment. Depending on the size distribution of the grains of sand and their random arrangement, we will actually experience a slightly different time interval every time we use the hourglass. This variation (or uncertainty) is considered the jitter in the hourglass. The resolution relates to the minimum time interval (seconds or fractions of a second, for example) that could be measured using the hourglass.

If we replace the grains of sand with electrons and the glass with a capacitor, then we have an electronic version of an hourglass that can measure time with much better accuracy and far better resolution. Figure 1.8 shows an ideal current source charging a capacitor. The event in this case is defined as the time it takes for the capacitor to charge from 0V (corresponding to an empty capacitor) to a threshold voltage, V_{TH} ($V_{DD}/2$, for example). If the current source is ideal, having an amplitude of I_0, this event takes CV_{TH}/I_0 to complete. Accordingly, this advanced hourglass has several knobs to turn in order to adjust its resolution. For example, one could reduce the capacitance, increase the current, or lower the threshold voltage of the comparator. The accuracy, however is limited by the current noise that is inherent in the methods by which we implement a current source.

A very simple current source may be implemented using a PMOS transistor with its gate connected to ground, as shown in Figure 1.9. This current source differs from an ideal current source in two ways: first, the current it produces is not constant; it becomes smaller as the capacitor is charged from 0 to V_{TH}; and second, the current includes noise such as thermal noise due to the random movements of electrons. The voltage dependency of the current slows down the process of charging the capacitor as there is simply less current. This deviation in time, however, can easily be absorbed by calibrating the time unit. The current noise, on the other hand, will result in different measurements of time as we repeat the experiment. To see this, let us model our practical current source as a constant current source in parallel with a noise current source, as shown in Figure 1.9. The current is now the sum of the nominal current, I_0, and the noise current. The additional noise current will cause the voltage across the capacitor to deviate from its ideal waveform, which is simply a ramp, and to arrive faster or slower at V_{TH}. This time deviation is a random process, which we call jitter, and is directly related to the characteristics of the noise current (itself a random process).

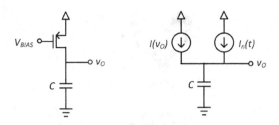

Figure 1.9 A simple PMOS current source charging a capacitor.

1.4   Jitter in a Ring Oscillator

The circuits we use to generate clock signals are called *oscillators*. One common type of oscillator, called a *ring oscillator*, consists of three inverters in a closed loop (a ring), as shown in Figure 1.1. We analyze this ring oscillator to explain intuitively how jitter is generated. A more rigorous treatment of this topic will be presented later in Chapter 4, once we mathematically define jitter and its characteristics.

As mentioned earlier in this chapter, the ring oscillator of Figure 1.1 produces oscillations (in voltage) with an expected period of $6t_{pd}$, where t_{pd} is the expected value of the propagation delay through each of the three stages. Let us now explore the properties of the delay of a CMOS inverter.

1.4.1   Jitter in Delay of a CMOS Inverter

When the input of an inverter rises from 0 to VDD, its output falls from VDD to 0. The high-to-low delay of an inverter (denoted by t_{phl}) is defined as the time elapsed between the input (at 50% of its swing) and the output (at 50% of its swing). The low-to-high delay of an inverter (denoted by t_{plh}) is defined similarly, as shown in Figure 1.10. If we assume the input transitions from low to high and high to low occur instantaneously, then t_{phl} and t_{plh} simply correspond to the time it takes for the output to reach 50% of its full swing. With this simplifying assumption, during the low-to-high transition of the output, only the PMOS transistor is ON, while during the high-to-low transition of the output, only the NMOS transistor is ON. Accordingly, we have two equivalent circuits to calculate t_{phl} and t_{plh}, as shown in Figure 1.11. In both circuits, two current sources charge or discharge the load capacitor. Of the two current sources in parallel, one is assumed to be deterministic and controlled by the gate voltage. The other source provides a random current corresponding to the thermal movement of electrons.

The deterministic parts of t_{phl} and t_{plh} are simply $C_L V_{TH}/I_0$. We refer to these as the *base delays* of the inverter. The remaining parts are two random variables which are also functions of time. We refer to these as the *excess delays*, or jitter, of the inverter. Jitter is random for the following reasons: if we measure two identical inverters at the same time, they exhibit different noise currents and hence result in different measurements of the delays. They are also time-dependent because if we attempt to measure t_{phl} (for

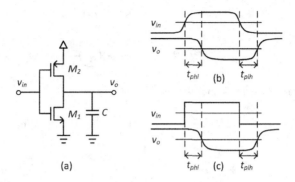

Figure 1.10 (a) A CMOS inverter, (b) propagation delays for non-ideal input, (c) propagation delays for ideal input.

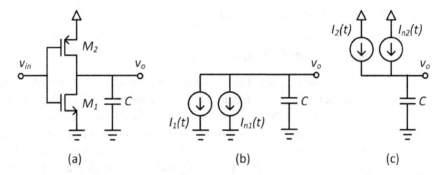

Figure 1.11 (a) A CMOS inverter, (b) the inverter model for high input, (c) the inverter model for low input.

example) of the same inverter, but at two different times, we end up with two different numbers, again due to the noise current changing with time. For these reasons, we say jitter is a random process.

To summarize, we have intuitively explained that the delay of a CMOS inverter has two components: the base delay and the excess delay, or jitter. The base delay is a deterministic number whereas the jitter is a random process (i.e., a time-dependent random variable). Accordingly, we can build a simple model of an inverter, as shown in Figure 1.12, where the input and the output are simply the time delays of the input and output signals with respect to a reference. Note that although we arrived at this model assuming an ideal (jitter-less) input, this model is still valid if the input has its own jitter. Since we are only concerned with jitter (i.e., the random part of any delay), we simply assume that both the input and the output of this model are random processes, where the output random process is the sum of the input random process and the random process created by the noise current inside the inverter. If we consider the input to the inverter to be an ideal clock, then the corresponding input to this model will simply be zero. We will use this simple model in the next section to better understand the jitter produced in a ring oscillator.

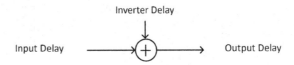

Figure 1.12 A simple model of an inverter.

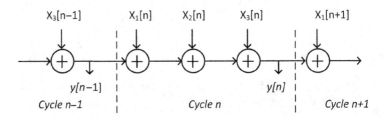

Figure 1.13 A simple model of a ring oscillator.

1.4.2   Modeling Jitter of the Ring Oscillator

Figure 1.13 shows a model to predict the absolute jitter of a three-stage ring oscillator. In this model, $X_i[n]$ represents the total excess delay introduced by inverter i in the n-th cycle, where n is the discrete time index. Note that this is the sum of the high-to-low and low-to-high excess delays of inverter i. The output (absolute) jitter is also described by a random process identified as $Y[n]$. Let us now explore some characteristics of $Y[n]$.

Given the model in Figure 1.13, one can clearly see that $Y[n]$ is the sum of the excess delays in each of the three inverters, plus the excess delay that was accumulated over all the past cycles. In other words, we can write: $Y[n] = Y[n-1] + X_1[n] + X_2[n] + X_3[n]$. If we further assume all $X_i[k]$s are identical ($X_i[k] = X$ for all i and k) and uncorrelated with each other, then $Y[n]$ would be a *random walk* process, which keeps a running sum of successive trials of a discrete-time random variable (X). It is well known [1] that the variance of a random walk process increases linearly with the number of trials, n. Similarly, the absolute jitter variance of a ring oscillator output grows linearly with the number of cycles, unless controlled by other means such as a phase-locked loop.

We can resort to the same model (Figure 1.13) to gain insight into the period jitter produced by the ring oscillator. Since period jitter is essentially the excess delay produced in a single cycle, it would be equal to the sum of the three excess delays only, that is $X_1[n] + X_2[n] + X_3[n]$; it does not account for the excess delays of previous cycles. As a result, the period jitter will have a finite variance, independent of the number of cycles.

1.5   Jitter in Electronic Systems

The timing uncertainty that manifests itself in charging a capacitor or in the delay of an inverter extends to all electronic systems, including digital circuits and systems, data converters, wireline, and wireless applications. However, the consequences or the

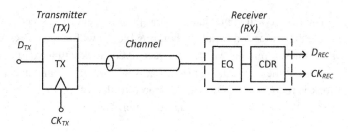

Figure 1.14 Block diagram of a wireline transceiver.

adverse effects of timing uncertainty are different among these applications. In digital systems, we are mostly concerned with completing an operation (e.g., multiplication) within a certain number of clock periods. In the presence of timing uncertainty, the time to completion will become uncertain. In a digital-to-analog converter, as we will see in Chapter 6, the timing uncertainty in the clock will result in voltage uncertainty at the analog output, compromising the accuracy of the analog voltage. We highlight the wireline application in this section to provide insight on how timing uncertainty can propagate through various blocks in a system and influence its performance.

A wireline link, as shown in Figure 1.14, consists of a transmitter, a channel, and a receiver. The transmitter, in its simplest form, sends digital "1" and "0" data on the rising edge of the transmit clock (CK_{TX}) to the channel. Since the transmit clock contains jitter, even when it is produced by a phase-locked loop, it will transfer this jitter to data edges. This is the first place in the link where the clock jitter is transferred directly to the data. Let us now follow the data through the channel and the receiver to see how this jitter is influenced by the channel and shaped by the receiver.

The channel is known to have a low-pass filter characteristic, caused mainly by the capacitive nature of the interconnect and the dielectric loss of the board. Equivalently, we can say that the channel has memory, i.e., the pulse corresponding to the previous bit (in a non-return-to-zero signaling) does not vanish immediately as we send the current bit, but partially interferes with the current and future bits. This interference, known as the inter-symbol interference (ISI), adds to the current bit level, and, as such, moves its zero crossings, creating jitter. This jitter is deterministic because it can be fully determined if the input data pattern and the channel response are both known. This is the second place in the link where jitter has been added to the data.

An equalizer (EQ) at the front end of the receiver is usually capable of reducing, or, in an ideal case, eliminating the deterministic jitter of the received data. However, the equalized data will still contain some random and some deterministic jitter. This equalized data is then fed to the Clock and Data Recovery (CDR) unit.

CDR is a simple feedback loop that controls the frequency and phase of an internal clock so as to minimize the phase error between the internal clock and the input data, producing the recovered clock. However, in this process, the recovered clock will inherit some of the input data jitter. In addition, other blocks in this control loop (such as a digital block that determines the phase difference between the input data and the recovered clock) will contribute additional jitter to the recovered clock.

The reader will appreciate that the jitter present in the recovered clock is influenced by the data jitter and by all the building blocks within the CDR. An immediate question that comes to mind is whether the recovered clock will have jitter characteristics similar to those of the data. Another question is whether the timing uncertainty in the recovered clock is reduced or increased compared to that of the data. To answer these questions, we need to characterize and quantify jitter. We will do this in Chapter 2.

2 Basics of Jitter

In the industry as well as in academia, the concept of jitter is sometimes treated using approximations, leading to misunderstandings and possibly to systems which are not optimally designed with respect to jitter performance. This chapter lays the foundation for a thorough and clear understanding of jitter in practice.

The chapter starts by providing four fundamental definitions of jitter: absolute jitter, relative jitter, period jitter and N-period jitter. It proceeds with an overview of other jitter definitions commonly found in literature. It will be shown that these additional definitions can be expressed in terms of the four fundamental jitter definitions.

A large part of the chapter handles the important topic of jitter statistics and the estimation of key parameters such as root mean square (RMS) and peak or peak-peak values. This will lead to the classification of jitter based on its distribution, and eventually to the introduction and explanation of the concepts of deterministic jitter, random jitter and total jitter.

2.1 General Jitter Terminology and Definitions

A variety of terms are commonly used to express the concept of jitter, and several different definitions can be found in the open literature, mostly dependent on the particular background of the author or on the specific application considered. A short and incomplete list of these terms includes period jitter, cycle jitter, cycle-to-cycle jitter, N-period jitter, accumulated jitter, adjacent jitter and long-term jitter. It is not uncommon to find the same term used with different meanings by different authors, leading to possible misunderstandings and confusion.

In this chapter jitter will be defined according to the way it can be measured, at least in principle. This will not result in jitter definitions that are different from the ones already found in the literature; rather, it will put the existing definitions in a clear and solid framework. These operative definitions of jitter turn out to be very intuitive, simple and of high practical value. Since the definition is connected to a measurement process, it is easier to relate it to a specific application. And, conversely, given a specific application, this approach makes it is easier to find the relevant jitter definition.

Generally speaking, jitter is the deviation of the time instant at which a given event occurs, relative to a reference time frame, which can be chosen arbitrarily. In the context of this book, the event we consider is the edge of a clock signal, or, more specifically, the time when a signal crosses a given threshold. The choice of the reference time frame can essentially be made in two ways: either the edges of the clock under investigation are compared to the edges of another clock, or they can be compared to some previous edges of the same clock (self-referenced). The first approach leads to the definition of absolute and relative jitter, while the second leads to the definition of period jitter. These three definitions of jitter, plus a fourth one, N-period jitter, which is an extension of the concept of period jitter, constitute the main topic of the next sections. We believe that most of the jitter aspects in modern electronic systems can be covered and properly described using only these definitions.

To set the stage for the next sections, let us introduce a generic clock signal $v(t)$ which will be used to clarify some initial concepts:

$$v(t) = A(t)\sin(\omega_0 t + \varphi(t)). \tag{2.1}$$

If we consider the quantity $A(t)$ as constant (independent of time t) and $\varphi(t)$ identical to zero, this signal is a perfect sinusoid with period $T = 2\pi/\omega_0$. The signal crosses zero with positive slope at equidistant times kT, with k any integer. We can consider this to be an ideal clock, in the sense that its zero crossings define a very accurate, noise-free time frame. Needless to say, this kind of signal does not exist in nature. In real applications, both $A(t)$ and $\varphi(t)$ are nonzero and depend on time, thus changing the characteristics of the signal.

The quantity $A(t)$ affects primarily the amplitude of the signal and is the origin of amplitude modulation, either intentionally, as in the case of AM data transmission, or unintentionally, in which case it is generally called amplitude noise or amplitude distortion. It is important to notice that, as long as $A(t)$ is not zero, the zero crossing of the signal are not perturbed and still occur at kT, so that this signal can still be used as ideal clock. As this condition is met for all practical clock signals, we will consider $A(t)$ to be constant, unless otherwise noted.

By contrast, the quantity $\varphi(t)$ added to the ideal phase $\omega_0 t$ in the argument of the sinusoid shifts the position of the zero crossings to deviate from the ideal instants kT, causing the phenomenon of jitter. The quantity φ is given different names in the literature: *excess phase*, *phase deviation*, *phase noise* or *phase jitter*. In order to avoid confusion, we will refer to it by the name *excess phase*. Note that the terms *phase noise* and *phase jitter* in particular might give rise to serious misunderstandings, as they are also used to describe different concepts (as will become clear in the next chapters), so they should be used with caution.

The signal shown in Equation 2.1 is just one particular case, in which the basic waveform is a sinusoid, but this is not the only possibility. Assume a generic waveform $x(\omega_0 t)$ periodic in t with period $T = 2\pi/\omega_0$ and with only one positive zero crossing per period. Such a waveform can be used to describe a generic clock signal as:

$$v(t) = A(t)x(\omega_0 t + \varphi(t)) \tag{2.2}$$

where the considerations above about $A(t)$ and $\varphi(t)$ can be exactly replicated.

In most of the practical applications in this book, the basic shape of the clock signal is rectangular. Although it is not common to talk about "phase" for a periodic rectangular signal, this concept can profitably be used if we think of the rectangular clock as the result of passing the signal 2.1 through a zero crossing comparator, with output A if the input is positive, $-A$ if negative and 0 if zero. This operation can be described with the help of the sgn function, so that the rectangular clock can be expressed as:

$$v(t) = A \operatorname{sgn}[\sin(\omega_0 t + \varphi(t))]. \tag{2.3}$$

With this in mind, all considerations about phase and excess phase for a sinusoidal signal can be transported to a rectangular signal.

2.1.1 Absolute Jitter

Assume the clock under investigation has a nominal period T. This means that the edges are affected by jitter, and each period of the clock is different, but the mean period is equal to T. One can think of comparing the position of each edge of the clock under investigation with the edges of another clock (called the *ideal clock* in the following) not affected by jitter and having exactly the same period T. The *absolute jitter* is defined as a discrete time random sequence **a**, where the k-th element, denoted as $\mathbf{a}_k$, is the time displacement of the k-th rising edge t_k of the real clock with respect to the corresponding edge of the ideal clock. This concept is explained in Figure 2.1.

Since the ideal clock is not jittery, its edges are spaced exactly by T. If the time axis is chosen properly, the edges of the ideal clock occur at time kT, so that the definition of absolute jitter can be given as:

$$\mathbf{a}_k := t_k - kT. \tag{2.4}$$

Since both the clock under investigation and the ideal clock have the same period, it is always possible to choose the position of the ideal clock so that the mean value of the absolute jitter is zero. If for any reason the position of the ideal clock is such that the average value of the absolute jitter calculated with Equation 2.4 is equal to a certain offset value $t_{OS} \neq 0$, the absolute jitter can be redefined as:

$$\mathbf{a}_k := t_k - kT - t_{OS}. \tag{2.5}$$

In the rest of the book, we will assume, unless otherwise stated, that the time axis is chosen so that $t_{OS} = 0$.

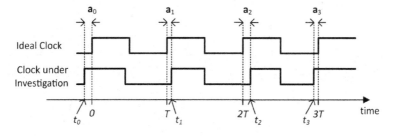

Figure 2.1 Illustration of the definition of absolute jitter.

The name "absolute" is chosen because the edges are compared to the edges of an ideal clock, which defines the time scale in an absolute manner, independent of any particular implementation or further reference. The ideal clock defines the absolute time scale to which the clock under investigation is compared.

2.1.2   Relative Jitter

In the previous section, the edges of the clock are compared to the edges of an ideal clock. However, one can think of comparing the edges of the clock under investigation (ck1) to the edges of another real, non-ideal clock (ck2), having the same average period T. This leads to the definition of *relative jitter* as a discrete-time random process $\mathbf{r}$, where the element $\mathbf{r}_k$ is the time displacement of the k-th rising edge t_k^{ck1} of the clock under investigation with respect to the corresponding edge t_k^{ck2} of the reference clock (see Figure 2.2):

$$\mathbf{r}_k := t_k^{ck1} - t_k^{ck2}. \tag{2.6}$$

It can easily be seen that the relative jitter can be expressed in terms of the absolute jitter of each of the two clocks considered. Indeed, adding and subtracting kT in the second term of Equation 2.6 and recalling Equation 2.4, the relative jitter can be rewritten as:

$$\mathbf{r}_k := \mathbf{a}_k^{ck1} - \mathbf{a}_k^{ck2}. \tag{2.7}$$

As in the case of absolute jitter, this definition assumes that the average value of the relative jitter is zero. If there is a fixed time offset between the edges of the two clocks, this offset must be subtracted, as is done for the absolute jitter in Equation 2.5.

2.1.3   Period Jitter

The two jitter definitions above are based on comparing the edges of the clock under investigation to the edges of another clock. Alternatively, one can think of comparing

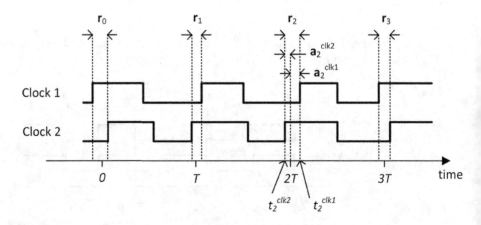

Figure 2.2 Illustration of the definition of relative jitter.

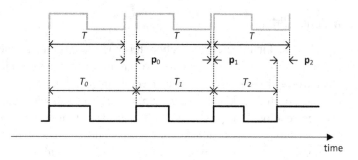

Figure 2.3 Illustration of the definition of period jitter. Each period of the clock is compared to the nominal period (gray clock signals).

the position of one edge of the clock with respect to the position of the previous edge of the same clock. Operatively, this corresponds to feeding the clock to an oscilloscope, triggering on one edge of the clock and looking at how much the following edge moves around its average position.

This procedure leads to the definition of *period jitter* as a discrete time random process **p**, where the element $\mathbf{p}_k$ is the variation around its nominal value of the position of one clock edge with respect to the previous edge. It is clear that this is equivalent to considering the variation of the actual clock period with respect to the nominal period. Figure 2.3 illustrates the concept. Each period of the clock is compared to the nominal period (gray clock signals). Any deviation contributes to period jitter.

The k-th sample of the period jitter will be indicated as $\mathbf{p}_k$ and can be mathematically defined as:

$$\mathbf{p}_k := (t_{k+1} - t_k) - T = T_k - T \tag{2.8}$$

where t_k and t_{k+1} are the time instants of two consecutive rising edges of the clock, T is the nominal clock period, and T_k is the actual clock period (see Figure 2.3).

By adding and subtracting $(k + 1)T$ to the right hand side of Equation 2.8, the period jitter can be expressed in terms of absolute jitter as:

$$\mathbf{p}_k = \mathbf{a}_{k+1} - \mathbf{a}_k. \tag{2.9}$$

In many cases it is of interest to know the variation of one edge relative not to the edge immediately preceding it, but to the N-th previous one. Consider for instance an ideal (delay-free and noiseless) digital frequency divider by N. The flip-flops in the divider are triggered by the rising edges of the incoming clock. Since the divider is noiseless and introduces no delay, the edges of one period of the output clock are perfectly aligned with the edges of the input clock spaced N periods apart. Therefore the period jitter of the output clock is equal to the relative variation of two input clock edges spaced N periods apart. This leads to the definition of N-period jitter, which is the subject of the next section.

2.1.4   N-Period Jitter

As mentioned in the previous section, it is possible to compare the position of one edge of the clock relative to the position of an arbitrary previous edge of the same clock. Operatively, this corresponds to connecting the clock to an oscilloscope, triggering on one edge of the clock and looking at how much the following edges move around their average position. This procedure leads to the definition of N-*period jitter* as the discrete time random process, denoted by $\mathbf{p}(N)$, where the element $\mathbf{p}_k(N)$ is the variation around the nominal value of the position of one clock edge with respect to the N-th previous edge.

This concept is illustrated in Figure 2.4 for $N = 5$, where a clock with nominal period T is shown affected by jitter. The k-th sample of the N-period jitter $\mathbf{p}_k(N)$ can then be expressed as the deviation of the time difference between the k-th and the $k+N$-th edges from the nominal value NT:

$$\mathbf{p}_k(N) := (t_{k+N} - t_k) - NT. \tag{2.10}$$

Considering that $t_{k+N} - t_k$ is equal to the duration of the first N periods of the clock, the expression for the N-period jitter can also be written as:

$$\mathbf{p}_k(N) = \left(\sum_{i=k}^{i=k+N-1} T_i \right) - NT \tag{2.11}$$

where T_i is used to indicate the i-th period of the clock.

By adding and subtracting $(k + N)T$ to the right side of Equation 2.10, the N-period jitter can be expressed in terms of absolute jitter as:

$$\mathbf{p}_k(N) = \mathbf{a}_{k+N} - \mathbf{a}_k. \tag{2.12}$$

Following from the definition above, it is clear that the value of N-period jitter for $N = 1$ corresponds to the period jitter defined in the previous section.

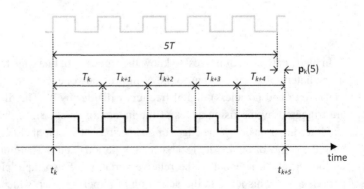

Figure 2.4 Illustration of the definition of N-period jitter.

An important and easily derived relation exists between N-period jitter and period jitter. Equation 2.11 can indeed be written as:

$$\mathbf{p}_k(N) = \sum_{i=k}^{i=k+N-1} (T_i - T) = \sum_{i=k}^{i=k+N-1} \mathbf{p}_i \qquad (2.13)$$

so that, recalling 2.8, the N-period jitter turns out to be equal to the sum of the period jitter over N consecutive periods. In the literature, the N-period jitter is also called *accumulated* jitter, since it originates from the accumulation of the jitter over consecutive periods. In this book, the term N-period jitter is preferred, but the two identify the same concept.

Before we proceed, one comment on notation: in a general context, the term N-period jitter refers to jitter over multiple periods, and N does not represent any particular number. When addressing a more specific case, though, N can be replaced by the actual number of periods being considered, or by the name of a variable or any other mathematical symbol representing it. One can thus speak of 5-period jitter, m-period jitter, or $(k+1)$-period jitter, with m and k representing integer numbers.

2.1.5   Other Jitter Definitions

In addition to the jitter definitions given in the previous sections, the technical literature often presents the reader with others. Some of them, like the Time Error (TE) or the Time Interval Error (TIE), are well known, standardized, and used extensively in specific applications. Others are used in papers, technical reports, application notes, and don't always share the same meaning.

This section will give a brief overview of the most popular among those additional jitter definitions. This list is by no means exhaustive or complete.

Time Error (TE) and Time Interval Error (TIE)

In 1996 the Telecommunication Standardization Sector of the International Telecommunication Union (ITU-T), the agency of the United Nations which regulates the interoperability of geographical communication networks worldwide, released the Recommendation G.810 [2]. This Recommendation standardizes definitions and terminology for telecommunication networks, among them the Time Error (TE) and the Time Interval Error (TIE).

Given a generic clock signal of the form reported in Equation 2.2, the Recommendation defines a *time function* $T(t) := \phi(t)/(2\pi f_0)$, where $\phi(t) := 2\pi f_0 t + \varphi(t)$ is the total phase of the clock signal under investigation and $\varphi(t)$ is the excess phase. Similarly it defines a *reference time function* $T_{ref}(t) := \phi_{ref}/(2\pi f_0)$ as the total phase of a reference clock signal divided by its radian frequency. Typically, an ideal clock with no excess phase (no jitter) is taken as reference signal, so that $T_{ref}(t) = t$.

After giving those definitions, the Recommendation defines the Time Error (TE) at time t as:

$$TE(t) := T(t) - T_{ref}(t). \qquad (2.14)$$

Based on the relations above and on the jitter definitions given in the previous sections, it is easy to show (this is left as an exercise for the reader) that the Time Error at a time t corresponding to the k-th edge of the clock under investigation is nothing other than the absolute jitter at k:

$$TE(t) = \mathbf{a}_k. \tag{2.15}$$

The Recommendation finally introduces the Time Interval Error (TIE) as a measure of the accuracy in determining the duration of a time interval τ when using the clock under investigation, as opposed to using the reference clock. It is defined as:

$$TIE(t, \tau) := [T(t + \tau) - T(t)] - [T_{ref}(t + \tau) - T_{ref}(t)]. \tag{2.16}$$

From here it can be shown that $TIE(t, \tau) = TE(t + \tau) - TE(t)$. Assuming that the times t and $t + \tau$ correspond to the k-th and $k + N$-th edges of the clock, respectively, using Equation 2.15 the Time Interval Error can be written as:

$$TIE(t, \tau) = \mathbf{a}_{k+N} - \mathbf{a}_k = \mathbf{p}_k(N). \tag{2.17}$$

It is thus revealed that the TIE over a time interval τ is nothing other than the N-period jitter, with N equal to the number of clock periods contained in τ.

A quantity connected to the TE and widely adopted by both the ITU-T and ANSI standardization bodies is the Maximum Time Interval Error, MTIE [3], defined as the maximum peak-to-peak value of TE(t) over a given interval τ:

$$MTIE_t(\tau) := \max_{t \leq t_1 \leq t+\tau} TE(t_1) - \min_{t \leq t_1 \leq t+\tau} TE(t_1). \tag{2.18}$$

The MTIE indicates the maximum error in the measure of any time interval in the time range from t to $t + \tau$ when using the clock under test. Following from the discussion above, the $MTIE_t(\tau)$ is simply equal to the peak-to-peak absolute jitter of the clock under test over the interval $[t, t + \tau]$.

Note that the name of Maximum Time Interval Error might be deceiving. Indeed $MTIE_t(\tau)$ is not necessarily identical to the maximum TIE over the time period from t to τ, as can be understood from the example in Figure 2.5.

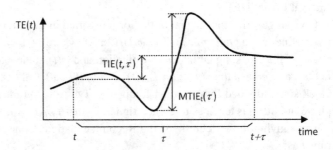

Figure 2.5 Example of TE versus time, an illustration of TIE and MTIE concepts.

Cycle-to-Cycle Jitter

The *cycle-to-cycle period jitter*, mostly known simply as *cycle-to-cycle jitter*, is defined in the JEDEC standard JESD65B [4] as the "variation in cycle time of a signal between adjacent cycles, over a random sample of adjacent cycle pair." It is construed as the difference between two consecutive clock periods and it indicates how much one period of the clock differs from the previous one. If we denote with T_k the k-th period of the clock, the cycle-to-cycle jitter $\mathbf{cc}_k$ is:

$$\mathbf{cc}_k := T_{k+1} - T_k \tag{2.19}$$

and can be easily expressed in terms of the period jitter as:

$$\mathbf{cc}_k = \mathbf{p}_{k+1} - \mathbf{p}_k \tag{2.20}$$

or in terms of absolute jitter as:

$$\mathbf{cc}_k = \mathbf{a}_{k+2} - 2\mathbf{a}_{k+1} + \mathbf{a}_k. \tag{2.21}$$

This jitter definition is mainly used to characterize the stability of spread spectrum clocks (SSC). To understand why this is a convenient approach, consider a triangularly modulated SSC where the profile of frequency versus time is a ramp. The excess phase, and thus the absolute jitter, is the integral of the frequency and follows a quadratic profile t^2. Since the period jitter is the first-order difference of the absolute jitter over time, it is also a function of time and changes as t. The cycle-to-cycle jitter, on the other hand, is the first-order difference of the period jitter over time and thus is ideally constant over one full ramp. This property makes the concept of cycle-to-cycle jitter more useful in the investigation of short-term jitter effects on SSC than either the period or the absolute jitter.

Note that, even though the term *cycle-to-cycle jitter* has been defined in standards, some authors have used it with a different meaning. For instance, before the release of the JEDEC JESD65B standard, the authors in [5] and [6] used this term to refer to the period jitter. It is worth mentioning that, in some contexts, the term *adjacent period jitter* is used instead of cycle-to-cycle jitter.

Long-Term Jitter

Although the term *long-term jitter* is widely used, it is not officially defined by any standardization body. Nevertheless, almost all sources agree and refer to it as the value of the N-period jitter for *large* values of N. The observation at the base of this definition is that for many practical clock generation circuits, like PLLs, the N-period jitter increases for small N, but then stabilizes around a value which is, in first approximation, independent of N. The reason for this will be explained in Section 3.2.2.

How large N should be to consider it "long term" depends both on the application and the clock generation unit. For instance, for a free-running oscillator, the N-period jitter increases indefinitely with N, so it doesn't make sense to speak of long-term jitter, since there is no asymptotic value for it. In PLLs the value of N for which the N-period jitter stops increasing depends on the PLL bandwidth (see Section 3.2.2). For a video system, where the line clock is synchronized at the beginning of one screen line, a large value of N might be of the order of the number of pixels in one line.

Short-Term Jitter

As with long-term jitter, the term *short-term jitter* has also not been standardized; however, unlike long-term jitter, different sources use it to mean different things. For instance, [7] refers to it as the cycle-to-cycle jitter, while [8] uses it as a synonym for period jitter. The reader is advised to take (and use) this term with caution and to make sure its exact meaning is clear before proceeding further.

Phase-Jitter or Integrated Jitter

The terms *phase-jitter* and *integrated jitter* have also not been standardized, but there seems to be a general consensus on their meanings. As in any other random process, the absolute jitter process can be decomposed into its frequency components and displayed in the frequency domain using its power spectral density as a function of frequency. The phase-jitter or integrated jitter is basically the RMS value of the absolute jitter, calculated considering only those frequency components of the power spectral density which fall within a given frequency interval. To investigate this definition further we need to introduce the concept of phase noise. This will be done in Section 3.1.5. It will then be clear that the terms "phase" in phase-jitter and "integrate" in integrated jitter come from the fact that this jitter is computed by integrating the phase noise over the desired frequency range.

Adjacent Period Jitter

This term is neither standard nor widespread. It is mostly used to indicate the cycle-to-cycle jitter (see, e.g., [9]).

Cycle Jitter

This term is sometimes used to indicate the jitter in one period or cycle of the clock (see, e.g., [10]). It is therefore equivalent to the period jitter.

Aperture Jitter

Aperture jitter is a term used specifically in the context of analog-to-digital conversion. In a sample and hold system, the time needed to disconnect the sampling capacitor from the input buffer by opening a switch is called the *aperture time*. This time is normally quite short, but not zero, and its effect can be modeled by introducing a small equivalent delay in the sampling instant of the analog waveform. The noise generated by the circuitry driving the switch introduces an uncertainty in the aperture time, which can be mapped to a variation in the small equivalent delay. This variation is what is called aperture jitter [11], [12]. Even though this type of jitter does not affect a clock or a data stream directly, its effect is equivalent to a jitter on the sampling clock of the data conversion system, which will be discussed in more depth in Chapter 6.

Table 2.1 Summary of fundamental jitter definitions.

Jitter	Symbol	Definition	vs. Edges' Timing
Absolute	$\mathbf{a}_k$	deviation of the clock edge position from the ideal one	$t_k - kT$
Relative	$\mathbf{r}_k$	difference of the corresponding edge positions of two clocks (A and B)	$t_k^A - t_k^B$
Period	$\mathbf{p}_k$	deviation of the clock period from its nominal value	$t_{k+1} - t_k - T$
N-period	$\mathbf{p}_k(N)$	deviation of the duration of N consecutive periods from its nominal value	$t_{k+N} - t_k - NT$
Cycle-to-cycle	$\mathbf{cc}_k$	difference between two consecutive clock periods	$t_{k+2} - 2t_{k+1} + t_k$

Table 2.2 Summary of fundamental jitter relationships.

Jitter	vs. Absolute Jitter	vs. Period Jitter
$\mathbf{a}_k$	–	N/A
$\mathbf{r}_k$	$\mathbf{a}_k^A - \mathbf{a}_k^B$	N/A
$\mathbf{p}_k$	$\mathbf{a}_{k+1} - \mathbf{a}_k$	–
$\mathbf{p}_k(N)$	$\mathbf{a}_{k+N} - \mathbf{a}_k$	$\sum_{i=k}^{k+N-1} \mathbf{p}_i$
$\mathbf{cc}_k$	$\mathbf{a}_{k+2} - 2\mathbf{a}_{k+1} + \mathbf{a}_k$	$\mathbf{p}_{k+1} - \mathbf{p}_k$

2.1.6 Summary of Jitter Definitions and Their Relationships

Tables 2.1 and 2.2 summarize the fundamental jitter definitions and their relations to absolute and period jitter.

2.2 Statistics on Jitter

In the previous sections, we defined different types of jitter. Independent of their specific definitions, all types are discrete time random processes. Absolute jitter, period jitter, and relative jitter are all functions of the time index k, the ordinal number indicating the clock edge which the jitter refers to. As such, each particular realization of a generic jitter process (that will be indicated with $\mathbf{j}_k$) can be represented on a graph with the index k as the horizontal axis and the value of $\mathbf{j}_k$ as the vertical axis, as shown in Figure 2.6.[1]

---

[1] Note that in this section the generic notation $\mathbf{j}_k$ is used instead of $\mathbf{a}_k$ or $\mathbf{p}_k$ or any of the other specific types of jitter, since the concepts explained in the following apply to all of them.

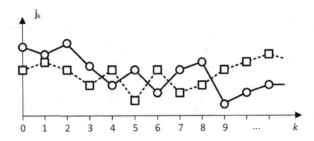

Figure 2.6 Two different realizations of the same jitter process.

The issue with this kind of representation is that the particular sequence $\mathbf{j}_k$, as, for instance, shown in Figure 2.6 with circles, is just *one* of the possible realizations of the jitter process. Measured at another time, the same clock will display a jitter sequence different from the first one, as shown in Figure 2.6 with squares.

It is therefore of fundamental importance to condense the information of the jitter process $\mathbf{j}_k$ into the smallest number of derived quantities which can capture all relevant information for any possible realization of the process $\mathbf{j}_k$, and be profitably used in system engineering.

One way to achieve this goal is to represent the sequence $\mathbf{j}_k$ by means of its spectrum, obtained via a Fourier transformation of its autocorrelation function, similar to what is done for deterministic signals.

The other way is to consider the distribution of the amplitudes of $\mathbf{j}_k$ on the y-axis by means of either a histogram, or, more accurately, the probability density function (PDF) and derive statistical data from it.

It must be noted that neither of these two approaches can fully replace the other. The spectral approach gives information on how the sequence $\mathbf{j}_k$ evolves over time, but is not able to capture very rare events accurately. For instance, a strong tone in the spectrum reveals the presence of a sinusoidal component in the sequence $\mathbf{j}_k$, but a value of $\mathbf{j}_k$ very far from the average and happening very infrequently (an outlier) will have almost no influence on the resulting spectrum and thus cannot be detected based on it. The statistical approach, on the other hand, while being capable of capturing outliers, is independent of the specific order of $\mathbf{j}_k$ over the index k, and thus cannot be used for inferring any behavior of the jitter process over time. As an example, Figure 2.7 shows two different sequences $\mathbf{j}_k$ leading to the same histogram (a uniform distribution between a and b), but having a completely different time behavior, and, therefore, different spectra.[2] It is clear that a particular system can react very differently, depending on whether the first or the second sequence is applied to it.

Both the frequency domain and the statistical approach therefore contain specific information. Different applications may benefit from one or the other, but only the combination of both gives a full picture of the jitter process.

This section will focus on the statistical approach while Chapter 3 will deal with the frequency domain approach.

[2] The reader should think of these sequences as deterministic signals for now. A more accurate definition of the spectrum of random signals is given in Appendix A.

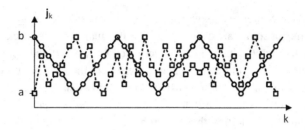

Figure 2.7 Two different jitter process having both a uniform distribution between *a* and *b*, but different time behaviors.

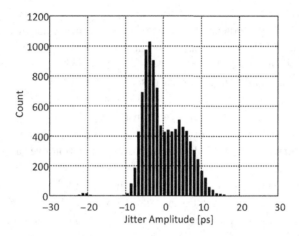

Figure 2.8 Generic jitter histogram.

2.2.1   Histograms and Probability Density Functions

The generic jitter process j_k is defined only for integer values of the index k, but the jitter amplitude – that is, the numerical value of the random variable j_k for a given k – can in principle assume any real value.

A common way to graphically represent the distribution of the amplitude of a random variable is a histogram. The jitter process j_k is measured and a number n of jitter samples are stored, corresponding to n different values of the index k. A histogram is then built by dividing the x-axis, representing the jitter amplitude, into a number of intervals of equal extension ("bins") and assigning to each bin a number equal to the number of jitter samples falling into that specific interval (the occurrence). Finally, the histogram is typically plotted as a bar chart, with the amplitude on the x-axis, and the occurrence in each bin on the y-axis. Figure 2.8 shows the histogram of a jitter process where ten thousand jitter samples were taken and binned into intervals of 1ps.

The histogram reveals a number of interesting properties of the jitter process under investigation. As an example, from the histogram in Figure 2.8 it can be clearly seen that the jitter amplitude is neither Gaussian nor symmetrical around its mean value – it shows two bumps where jitter amplitude seems to be more frequent – and, finally, there are a few jitter events which fall on the left side far away from the main distribution, hinting at some rare event corrupting the quality of the clock.

Although the histogram is a very well known and widespread graphical tool, it is, for many reasons, more convenient to represent the distribution of jitter amplitude (or any other random variable) by using a probability density function (PDF). For a continuous random variable x the PDF is a real function of the possible values assumed by x, satisfying the condition that the probability of x assuming values in any interval $[a, b]$ is equal to the integral of the PDF from a to b:

$$P[a < x < b] = \int_a^b f_x(c)dc \qquad (2.22)$$

where $P[U]$ indicates the probability of event U, and $f_x(.)$ is the PDF of the random variable x. From this basic condition, it follows that the PDF is always positive or zero and its integral from $-\infty$ to $+\infty$ is equal to one.

At this point an important observation on a fundamental difference between a histogram and the PDF must be considered. A histogram is derived from the data of one particular realization of a random process, measured over time. For each instant of time only one value of the process is recorded, and the statistics are built by recording many samples of the same process realization over a given time span. A PDF, on the other hand, is the description of the distribution of the values assumed by random processes at one particular time instant. A PDF is equivalent to generating many parallel realizations of the same jitter process and considering the jitter value at some specific moment in time (a possibility that is, however, normally not given in practice). In other words, the PDF is a property of one point in time over all possible realizations, while the histogram is a property of one single realization over all instants of time. The usual assumption is that the two descriptions are equivalent; that is, that temporal averages over one realization give the same result as averages over the ensemble of realizations. This assumption, although reasonable, is not always true. Processes for which this assumption holds are called *ergodic* and are a subset of stationary processes [1] (see also A.2.11 for a short review of ergodicity). Fortunately, in most of the processes where jitter is involved, the ergodicity is given, so that we can use histograms to infer statistical properties of the process in one specific point in time. In the following discussion, we will assume that the processes we are dealing with are ergodic.

Once the histogram of a random process is available, it is easy to derive the corresponding PDF.

Assume the amplitude histogram (h_j) of a jitter process j_k is built on m bins of extension Δt, centered around the values t_i, with $i = 1 \ldots m$. Then, the value $h_j(t_i)$ of the histogram for the i-th bin is the number of jitter samples falling in the interval $[t_i - \Delta t/2, t_i + \Delta t/2]$. From the way the histogram is built, the sum of all $h_j(t_i)$ is equal to the total number of jitter samples n:

$$\sum_{i=1}^m h_j(t_i) = n. \qquad (2.23)$$

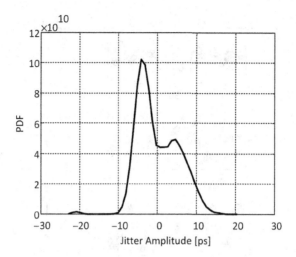

Figure 2.9 Generic jitter probability density function.

If we define a function $f_j(t_i)$ as the histogram scaled by the factor $1/(n\Delta t)$:

$$f_j(t_i) := \frac{h_j(t_i)}{n\Delta t} \qquad (2.24)$$

by virtue of Equation 2.23 the following result can be obtained:

$$\sum_{i=1}^{m} f_j(t_i)\Delta t = 1. \qquad (2.25)$$

The left-hand side of this equation can be interpreted as the discrete version of the integral of the function $f_j(t)$ over the whole t-axis. Therefore the integral of the function defined in Equation 2.24 sums to 1. It is left to the reader to prove that the integral of $f_j(t)$ between any two values a and b is equal to the number of jitter samples falling between a and b divided by the total number of samples, and thus equals the probability $P[a < j_k < b]$. The function $f_j(t)$ defined as in Equation 2.24 is thus the PDF of the jitter process j_k. Figure 2.9 shows the PDF of the jitter amplitude corresponding to the histogram in Figure 2.8. As can be seen, moving from histogram to PDF implies only a scaling of the y-axis, but no change in shape. Note also that, since the extension of the PDF on the x-axis is normally in the range of picoseconds (10^{-12}), the value of the PDF of the y-axis is in the range of 10^{+12}.

In practice, jitter in circuits can be caused by many different mechanisms and the PDF can assume different shapes. Figure 2.10 illustrates some sample realizations of the most typical jitter processes found in electronics and their PDFs: Gaussian, uniform, sinusoidal and Dirac. It can be seen how each process shows a typical signature in the PDF. In general the opposite is not true, since a PDF shape can, in principle, be generated by sequences with very different time domain behaviors. However, in many practical cases, the shape of the PDF can give a very good hint about the underlying time domain sequence, if ergodicity is given. It must be noted that the

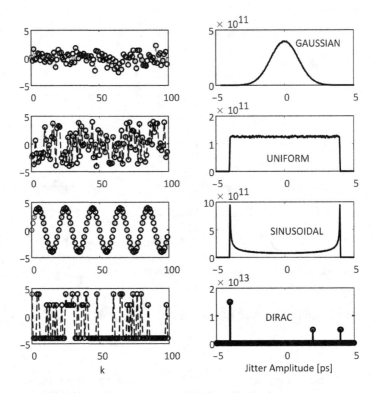

Figure 2.10 Probability density functions for typical jitter distributions.

Dirac distribution is rarely found in practical cases, but it is a very useful abstraction for cases in which the jitter is concentrated around two or more discrete values. The Dirac distribution with only two discrete values, also called *Dual Dirac*, plays an important role in jitter decomposition, as will be seen in more detail in the next sections.

For completeness, the expressions of these most common PDFs are reported here. For a Gaussian jitter with RMS value σ and average value μ the PDF is given by:

$$f_{\mathbf{j}}(t) = \frac{1}{\sigma\sqrt{2\pi}} \exp\left(\frac{-(t-\mu)^2}{2\sigma^2}\right). \tag{2.26}$$

For a uniform jitter between two values A and B, the PDF is simply given by:

$$f_{\mathbf{j}}(t) = \frac{1}{B-A} \tag{2.27}$$

for $A \leq t \leq B$, and zero otherwise. For a sinusoidal jitter with peak amplitude A and average value μ the PDF is:

$$f_{\mathbf{j}}(t) = \frac{1}{\pi A} \frac{1}{\sqrt{1-((t-\mu)/A)^2}} \tag{2.28}$$

for $\mu - A \leq t \leq \mu + A$ and zero otherwise. Finally, for a Dirac jitter, assuming m discrete values t_i, with $i = 1 \ldots m$, the PDF can be written as:

$$f_j(t) = \sum_{i=1}^{m} A_i \delta(t - t_i) \qquad (2.29)$$

where $\sum A_i = 1$ to ensure that the integral of the PDF sums to 1.

Each of the mechanisms generating jitter according to the PDFs described above rarely occurs in isolation. In nature, multiple mechanism are at work simultaneously, so that the PDF of the resulting jitter can assume more complicated shapes, as shown as an example in Figure 2.9. Section 2.2.7 will explain how different PDFs combine into one single resulting PDF.

Although the shape of the PDF already provides a great deal of information about the jitter process, it is very useful to extract some basic statistical parameters from it. In industrial contexts, the most commonly used parameters for describing jitter are: the mean, the median, the variance and its square root, the standard deviation (also called RMS, root mean square), the peak and the peak-peak. The next sections will go into the details of each of them.

2.2.2   Jitter Mean

In many cases jitter is defined such that its mean is equal to zero. As an example, the period jitter as defined in Equation 2.8 is intrinsically a zero mean process. In such cases a nonzero mean value would imply that the nominal period T has been calculated incorrectly. Absolute jitter is also another example where, if correctly extracted, the mean value is intrinsically zero. However, there are cases where the mean might be nonzero. One such case is the relative jitter as defined in Equation 2.6. If the two clocks have a static timing skew between them, the average of the jitter is equal to the timing skew.

When needed, the mean μ_j of a generic jitter process $\mathbf{j}_k$ can be estimated based on n samples by the so called *sample mean*:

$$\hat{\mu}_j = \frac{1}{n} \sum_{k=1}^{n} \mathbf{j}_k \qquad (2.30)$$

where the hat symbol indicates that this is an estimation.

2.2.3   Jitter Median

Similar to the mean, another measure of the central tendency of the jitter distribution is the median. The median is defined as that particular value j such that 50% of the jitter samples are smaller and 50% are larger than j, or, equivalently, a particular jitter sample falls with 50% probability above j (and obviously also with 50% probability below it):

$$P[\mathbf{j}_k < j] = P[\mathbf{j}_k > j] = 0.5. \qquad (2.31)$$

If the jitter PDF is symmetric, then the mean and median coincide, but in many cases the PDF is skewed (e.g., there is a longer tail on one side than on the other) and the mean and median are not the same.

Although the mean is the most commonly used measure of central tendency, there are applications where the jitter median plays an important role. One notable example is in CDRs (or PLLs) with binary phase detectors. In these systems the loop locks around the median value of the relative jitter between data (or reference clock) and feedback clock, not around its mean value. If the relative jitter distribution is very asymmetric, the difference in locking point can be significant and close the eye more than would be expected if the mean were considered.

2.2.4   Jitter Standard Deviation and Variance

One of the most widespread statistical parameters for characterizing jitter is the root mean square (RMS) value. In statistics, the RMS is known as standard deviation, indicated as σ, and is the square root of the variance σ^2. In this book we use the terms RMS and standard deviation as synonyms.

The standard deviation is the main parameter for the characterization of Gaussian distributions, but it can be computed for any kind of distribution. It is a measure of how wide the jitter spreads around its average value, and plays an important role in the design of many electronic systems.

The usual way of estimating the RMS value is first to estimate the variance of the jitter and then extract its square root. The variance of a generic jitter process $\mathbf{j}_k$ can be estimated based on n samples by the so-called *sample variance*:

$$\hat{\sigma}_{\mathbf{j}}^2 = \frac{1}{n-1} \sum_{k=1}^{n} (\mathbf{j}_k - \mu_{\mathbf{j}})^2 \tag{2.32}$$

where $\mu_{\mathbf{j}}$ is the mean of the process $\mathbf{j}_k$ and the hat symbol indicates again that this is an estimated value (see, e.g., [13]).

2.2.5   Jitter Peak and Peak-Peak

Many applications are affected by the extreme values that jitter can assume rather than by the RMS value of jitter. For example, the error-free operation of a synchronous digital circuit is guaranteed if the minimum duration of the clock period is still large enough for the combinatorial logic to produce its output sufficiently in advance of the next clock edge. In this case, the quantity that has to be characterized is the minimum duration of the period, or, in other terms, the maximum value for the period jitter (Chapter 5 will deal with this application case in detail).

In these cases, we speak of *peak* and *peak-peak* as a measure of the maximum and maximum minus minimum values that jitter can assume. It should be noted that these measures are mostly meaningless if they are not accompanied by additional information about how they were performed. To understand this, imagine the very common case of a jitter process with a Gaussian distribution. If 100 samples are measured, a histogram with a given extension is obtained. Due to the unboundedness of the Gaussian distribution, though, it is very likely that if one thousand samples of the same process are measured, the resulting histogram range, and therefore the peak or peak-peak values, will be larger than in the case using a smaller sample size. The peak and peak-peak jitter therefore must be defined more precisely.

The underlying concern when looking for the peak or peak-peak values is to determine the worst-case scenario, the worst jitter value that can be expected. Since circuits are affected by thermal noise and thermal noise is Gaussian, the peak jitter should always be infinite from a purely mathematical point of view. Of course, no system would work if this were true, so there must be a way out of this dilemma. One approach is to ask the following question: once a number for the peak value is given, what is the probability that some jitter samples will be larger that this number? Or, alternatively, how can we find a number such that the probability that jitter samples exceed this number is lower than a predefined bound?

There are two ways of answering this question. The first, and probably the most common, does not make any assumptions about the jitter distribution, while the second leverages the knowledge of the distribution shape.

The first way is to define peak $\rho_{\mathbf{j}}$ and peak-peak $\rho\rho_{\mathbf{j}}$ jitter by taking n jitter samples $\mathbf{j}_k$, $k = 1 \dots n$ and finding the maximum and minimum values among the samples:

$$\rho_{\mathbf{j}}^{+} := \max_k(\mathbf{j}_k) - \mu_{\mathbf{j}} \tag{2.33}$$

$$\rho_{\mathbf{j}}^{-} := \mu_{\mathbf{j}} - \min_k(\mathbf{j}_k) \tag{2.34}$$

$$\rho\rho_{\mathbf{j}} := \max_k(\mathbf{j}_k) - \min_k(\mathbf{j}_k) \tag{2.35}$$

where $\mu_{\mathbf{j}}$ is the jitter mean, as defined in Equation 2.30. Note that, for the peak values, since some distribution might be asymmetric around the mean value, it is necessary to distinguish between *peak positive* (ρ^{+}), the range of the jitter on the right-hand side of the mean, and *peak negative* (ρ^{-}), the range of the jitter on the left-hand side of the mean. Figure 2.11, illustrates these definitions with an example histogram.

It will be shown in Section 9.2.3 how the probability of finding jitter samples outside the peak bounds defined in this way is essentially inversely proportional to the sample size. For some applications, very small probabilities (of the order of 10^{-12} or lower)

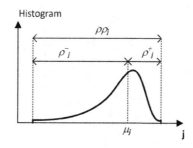

Figure 2.11 Definition of peak positive jitter $\rho_{\mathbf{j}}^{+}$, peak negative jitter $\rho_{\mathbf{j}}^{-}$ and peak-peak jitter $\rho\rho_{\mathbf{j}}$.

are specified, requesting the collection and processing of a prohibitively large amount of data. Another issue connected with this procedure of finding the peak values is that statistics based on finding the maximum (or minimum) values from a population are the *least* robust. It can be understood that a single outlier in the set of measurements $\mathbf{j}_k$ – due to, e.g., a single measurement error or a glitch – can corrupt the whole result. Therefore, the data set must be perfectly clean and free from erroneous outliers, before applying the above-mentioned method.

For these reasons, a second, more robust approach is needed, which leverages prior knowledge of the jitter distribution. Assuming that we have reason to believe that the jitter distribution is Gaussian, or very close to Gaussian, there is a well-defined probability of finding a jitter sample deviating from the mean more than Q times the RMS value $\sigma_\mathbf{j}$. The reader is referred to Section 9.2 for details about how to compute this probability. Since the RMS value can be estimated by Equation 2.32, the peak (in this case, peak positive and peak negative are the same) and peak-peak values can be defined as:

$$\rho_\mathbf{j} := Q \cdot \sigma_\mathbf{j} \tag{2.36}$$

$$\rho\rho_\mathbf{j} := 2Q \cdot \sigma_\mathbf{j}. \tag{2.37}$$

Using these definitions, the probability of finding jitter samples outside the peak or peak-peak ranges are $P(Q)$ and $2P(Q)$ respectively, with $P(Q)$ defined in Equation 9.14 and tabulated in Table 9.1 Note that these definitions allow us to find the peak values for very low probabilities without the need to process huge amounts of data, and a few outliers have a small impact on the estimation of the RMS, even for moderate sample sizes.

But what if the jitter distribution is non-Gaussian, as often happens in real applications? In this case, the estimated RMS cannot be correctly used to compute the marginal probability of jitter outside given bounds. To illustrate this concept, Figure 2.12 shows on the left-hand side the PDF of a real jitter distribution (solid line). The estimated RMS using Formula 2.32 is 2.204ps and the dashed line shows a Gaussian distribution with exactly this RMS value. With reference to these two curves, the right-hand side plots the

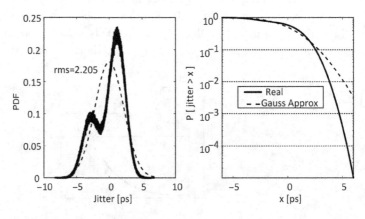

Figure 2.12 Example of non-Gaussian jitter distribution (left) and its corresponding probability curve (right).

probability of finding a jitter sample larger than x (the x-axis) for the real case (solid line) and for the Gaussian approximation(dashed line). It is evident that predictions based on the Gaussian approximation would be wrong.

For these reasons, a different methodology has been developed to extract meaningful statistical values from distributions which are not Gaussian. This methodology requires knowledge about deterministic jitter, which will be the topic of the next sections.

2.2.6   Taxonomy of Jitter Based on PDF

The methodology developed to cope with non-Gaussian jitter distributions is built on the idea that the jitter on a clock, or data, in a complex system is the combined result of multiple different mechanisms, each of which contributes one component of the resulting jitter. For instance, as will be seen in Section 1.5, the jitter on data at the receiver front of a wireline system end is the combination of the jitter produced by the transmit PLL, plus the jitter introduced by the thermal noise of the off-chip driver, plus the distortion of the edges due to the band limitation of the channel itself, among others.

One erroneous bias is assuming that jitter is basically *random* only. This is not true; jitter can manifest itself in a random as well as in a deterministic fashion. In analogy to the more familiar case of amplitude modulated signals, deterministic jitter can be thought of as the "horizontal" equivalent of distortion or cross-talk effects on the "vertical" axis of an analog signal, while random jitter can be thought of as the equivalent of the classical Gaussian amplitude noise.

More specifically, each jitter component encountered in real applications can be classified in one of two categories, based on the shape of its histogram or PDF: *Deterministic Jitter* (DJ) and *Random Jitter* (RJ). Note that this classification can be applied to any type of jitter as presented in the previous sections (period jitter, N-period jitter, absolute jitter, ...).

Random Jitter (RJ)

The RJ includes all jitter components whose PDFs are *unbounded*, meaning that their range grows indefinitely the more jitter samples are considered. In electronic systems, RJ is produced by the electronic noise of the devices (thermal, flicker, and shot), and manifests itself as a jitter with a Gaussian distribution.

It must be noted that the term RJ is generally used not only to describe the unbounded nature of jitter, but also as a numerical value indicating the amplitude of the random jitter component itself. There is no unified convention so far, though, so different standards use the term RJ differently. For instance, in Serial-ATA, the RJ identifies the RMS value multiplied by a factor which depends on the specified bit error rate (BER), e.g., for a BER $= 10^{-12}$, the factor is 14, as will be explained in the next sections. The PCIe standard, on the other hand, uses RJ to indicate the RMS of the Gaussian distribution, without multiplying it by the factor depending on the BER. In this book, we will stick with the first convention, as it seems to be the most widely used. As an example, for a clock affected by Gaussian jitter with RMS 1ps, one can say that the clock has a random jitter component and its RJ is equal to 14ps, if a probability of error of 10^{-12} is targeted.

Deterministic Jitter (DJ)

The DJ category includes all jitter components whose PDF is bounded. The most common causes of DJ are modulation of the clock due to supply noise, cross-talk from other signals or channels, duty cycle distortion, channel bandwidth limitation, and the like. Depending on the originating mechanism, the DJ can be divided in several subcategories, the most important being:

- data-dependent jitter (DDJ): specific to jitter on serial digital data, it includes jitter effects which are correlated with the transmitted data.
- duty-cycle-distortion jitter (DCD): jitter due to asymmetries in the duty cycle when both rising and falling edges of the clock are used in the application.
- bounded-uncorrelated jitter (BUJ): jitter specifically in serial digital data, which is bounded but bears no correlation with the transmitted data. It may be due to, e.g., crosstalk from adjacent channels and can be subdivided in periodic jitter (PJ) and non-periodic jitter.
- sinusoidal jitter (SJ): jitter with a sinusoidal profile, usually used to test jitter tolerance in high-speed interfaces.

Additionally, in the case of serial data transmission, jitter can also be divided into Correlated and Uncorrelated, depending on its relation to the transmitted data. There is no standardized naming of those subcategories, and different authors often use different terms to refer to the same concept. However the basic distinction to keep in mind is between RJ, which is unbounded and mostly Gaussian distributed, and DJ, which is bounded, with distributions of different shapes. Figure 2.13 illustrates this taxonomic scheme. For an exhaustive list of the sub-categories see, e.g., [14].

It must be noted that the term DJ is generally used not only to describe the bounded nature of jitter, but also as a numerical value indicating the peak-peak value of the jitter component itself. As an example, for a clock affected by sinusoidally modulated jitter with an amplitude of 1ps, the clock has a deterministic jitter component with a DJ equal to 2ps.

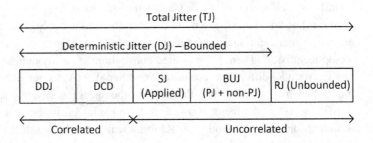

Figure 2.13 Taxonomy of jitter terminology and their relationships.

2.2.7 Combination of Jitter Components

The topic to investigate next is how the different jitter components combine to generate the resulting jitter. As an example, take the wireline system described in Figure 1.14. The data transitions at the input of the TX driver are already affected by jitter, mainly produced by the PLL. The jitter introduced by the thermal noise in the driver shifts the edges of the output data in a manner that is largely independent from how much jitter is already present on them. In other words, the displacement of the data edges produced by the noise of the driver is *added* to the displacement already present on the data at the input. Generalizing this concept, it can be said that, in most cases, the resulting jitter $\mathbf{j}_k$ can be expressed as the sum of the several jitter components $\mathbf{j}_{1,k}, \ldots, \mathbf{j}_{n,k}$:

$$\mathbf{j}_k = \mathbf{j}_{1,k} + \mathbf{j}_{2,k} + \ldots + \mathbf{j}_{n,k}. \tag{2.38}$$

It must be noted that the assumption that the amount of jitter introduced by each mechanism is independent from the jitter due to other mechanisms is not always true, and, in some cases, the jitter already present on the clock does influence the amount of jitter that another specific noise mechanism adds to the clock. However, in most cases, Equation 2.38 holds.

Assuming that the jitter components are independent, a basic result of probability theory states that the PDF of the resulting jitter, $f_{\mathbf{j}}(t)$, is the convolution of the PDF of the single components:

$$f_{\mathbf{j}}(t) = f_{\mathbf{j}1}(t) * f_{\mathbf{j}2}(t) * \ldots * f_{\mathbf{j}n}(t). \tag{2.39}$$

It is very instructive to take a look at how the convolution process changes the shape of the resulting jitter PDF in some exemplary cases. Figure 2.14 shows the result of the convolution of a Gaussian distribution (RMS = 0.2) with DJs having a dual Dirac, a sinusoidal, a uniform, and finally a triangular distribution, from top to bottom, respectively. The original DJ distributions are shown with a dashed line, while the resulting distributions are shown with a solid line. Note that the range of the bounded distributions is from -1 to $+1$ for all of them, meaning that their peak-peak value is 2. Note also that for the sake of graphical clarity, the distributions have been normalized vertically to a maximum value of one.

It is evident how the convolution process smooths out the hard borders of the DJ distributions. It can be seen that the positions of the peaks, if there are any, or of the maximum levels reached by the resulting distribution do not coincide with the positions of the extremes of DJ; they are instead moved inwards. The only exception is when the DJ is an ideal dual Dirac distribution. This shift inwards is more pronounced the more tapered the DJ is. It is therefore not easy to determine the amplitude of the DJ by visual inspection of the resulting PDF, as is evident in the case of the triangular DJ.

The Central Limit Theorem of probability states that the sum of multiple independent random variables tends to resemble a Gaussian distribution. Thus, when multiple DJ components are present, the resulting DJ will have a shape that is very close to a Gaussian curve, with the only difference being that its tails will be bounded. Since the convolution of two Gaussian distributions is still a Gaussian distribution, in this

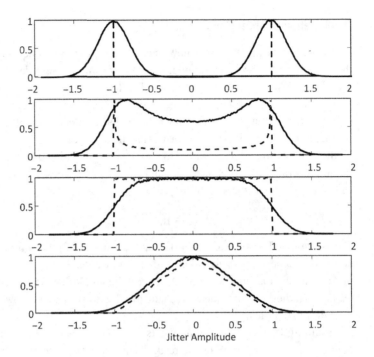

Figure 2.14 The solid lines are the PDF of the sum of Gaussian jitter (RMS=0.2) and DJ components (DJ=2) with different distributions, shown dashed (from top to bottom: dual Dirac, sinusoidal, uniform, and triangular). The distributions have been normalized vertically to a maximum value of one for graphical clarity.

particular case the distribution of the sum of RJ and DJ will have a shape very close to a Gaussian curve, and determining the DJ by visual inspection is simply not possible.

From Equations 2.38 and 2.39, it can be understood that when multiple uncorrelated RJ or DJ components are present, the total RJ has a variance which is the sum of the variances of the single RJ components:

$$\sigma_{total}^2 = \sigma_{j1}^2 + \sigma_{j2}^2 + \cdots + \sigma_{jn}^2 \tag{2.40}$$

while the total DJ is the linear sum of the single DJ components:

$$DJ_{total} = DJ_{j1} + DJ_{j2} + \cdots + DJ_{jn}. \tag{2.41}$$

The first equation above can also be derived from the fact that the sum of two uncorrelated Gaussian random variables is still Gaussian, with a variance given by the sum of the variances. The second comes from the known property of the convolution of two functions with bounded ranges: the range of the result is the sum of the ranges of the convolved functions. In short, RJ adds quadratically while DJ adds linearly.

Since single RJ components in a system are the result of multiple thermal noise processes which are by nature uncorrelated to each other, Equation 2.40 gives a very good estimation of the total RJ.

For DJ, though, this assumption is not always true. To clarify this point, consider a clock signal going through two cascaded buffering stages connected to two different noisy supplies. Assume now that the noise on each supply is sinusoidal with a given frequency (the same for the two supplies), but with 90 degrees phase shift between the two. In this case when the noise on the first buffer stage is at its maximum, the noise on the second one is zero. It is clear that the DJ at the output of the buffer chain is less than the arithmetic sum of the DJ produced by each of the supplies on the two buffering stages separately. In general, if DJ components show a strong correlation the total DJ can not be computed using Equation 2.39, and its range is less than, or, at most, equal to the sum of the ranges of its single components. Therefore Equation 2.41 is an upper bound to the total DJ, and can be taken as a worst case scenario when designing the system. In some cases, though, simply summing linearly the DJ could lead to a very pessimistic estimation and to an overly conservative designed system. In such cases, a more in-depth analysis of the sources of DJ and their correlations must be done.

2.2.8 Jitter Decomposition

With the insight gained in the previous sections, we now go back to the original question of how to derive meaningful statistical parameters from a jitter PDF which is not Gaussian. Jitter decomposition is a methodology developed to tackle this problem. The basic idea is to separate (decompose) the original jitter PDF into its RJ and DJ components and derive the jitter statistics based on these two values.

There are several methods developed for this purpose. The two most popular among them are the so-called *independent-σ technique* and the *tail fitting*.

The independent-σ technique relies on measurement in the frequency domain, specifically on the phase noise of the clock under investigation, to extract the RMS value of the RJ component, as will be seen in Chapter 3. With this information available, the DJ is extracted either by deconvolution or by trying to fit the tails of the original distribution with a Gaussian curve having an RMS value equal to the RJ and variable mean values.

The second technique, tail fitting, is far more popular. Since the jitter PDF is the result of a convolution of a bounded DJ with a Gaussian RJ, it can be expected that the tails of the jitter PDF are still very close to a Gaussian shape, especially for offsets very far from the bounds of the DJ distribution. It therefore makes sense to try to fit the left and right tails of the jitter distribution with Gaussian curves parametrized in terms of mean μ, RMS value σ and amplitude A. The amplitude parameter A is a factor multiplying the standard Gaussian PDF and is needed since the left- and right-fitted Gaussian curves account only for a fraction of the whole PDF and thus have an area smaller than one. The tail fitting is an attempt to find the best A, σ and μ fitting the left and right tails of the original jitter distribution.

In a simplified approach, the amplitude A is assumed constant and equal to one. Although easier from an algorithmic perspective, it tends to underestimate the DJ component and overestimate the RJ one, as will be shown in the next section. Thus the predictions on the probability of error using this approach are not very accurate.

A more advanced approach considers also the amplitude A as a optimization parameter. The result is six parameters: A_L, σ_L, and μ_L for the left tail and A_R, σ_R, and μ_R for the right tail. Figure 2.15 shows the result of the tail fitting applied to the distribution of

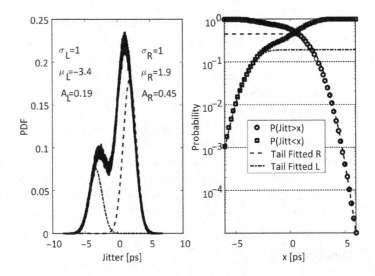

Figure 2.15 Non-Gaussian jitter distribution with Gaussian tail fitting (left) and corresponding probability curves (right). Left hand side: the original jitter PDF (solid line) and the two fitted Gaussian curves (dashed lines). Right hand side: probability of finding a jitter sample larger or smaller than a given value *x*, based on the real jitter PDF (markers) and based on the fitted Gaussians (lines).

Figure 2.12. It can be seen that tail fitting provides a very good description of the tail behavior in terms of probability. Note also how this approach focuses only on the behavior of the extremes of the jitter distribution, and does not attempt to describe accurately the central part, which is normally DJ-dominated. For most applications, this fact does not represent a real limitation, since the critical part of the jitter PDF lies in the tails.

While the parameters A and μ can be different for the left and for the right tails, σ_L and σ_R should typically converge to the same numerical value for a real electronic system.

Note how the tail fitting approach applies a *Dual-Dirac model* to the PDF under investigation, since it essentially approximates the original PDF with the convolution of a Gaussian jitter with a Dual-Dirac jitter, where the two Dirac impulses are centered at μ_L and μ_R and have amplitudes A_L and A_R.

2.2.9   Total Jitter and Probability of Error

Based on the parameters derived from the tail fitting procedure, the values of peak left, peak right and peak-peak can be defined for any jitter distribution as:

$$\rho^- = \mu_L - Q \cdot \sigma_L \tag{2.42}$$

$$\rho^+ = \mu_R + Q \cdot \sigma_R \tag{2.43}$$

$$\rho\rho = \mu_R - \mu_L + Q(\sigma_R + \sigma_L) \tag{2.44}$$

where Q is a factor that determines the probability of finding a jitter sample outside of the so-defined extremes. Note that, since the fitted Gaussians are weighted by the scale factors A_L and A_R, the probability of finding jitter higher than ρ^+, lower than ρ^-, or outside $\rho\rho$ are:

$$P\left[\mathbf{j}_k > \rho^+\right] = A_R \cdot P(Q) \tag{2.45}$$

$$P\left[\mathbf{j}_k < \rho^-\right] = A_L \cdot P(Q) \tag{2.46}$$

$$P\left[\mathbf{j}_k \text{ outside } \rho\rho\right] = (A_R + A_L) \cdot P(Q) \tag{2.47}$$

where $P(Q)$ is the expression reported in Equation 9.14 and tabulated in Table 9.1.

In Equation 2.44 the quantity $\mu_R - \mu_L$ accounts for the effect of the DJ components on the jitter distribution and is often referred to as *Dual-Dirac DJ* and indicated with $DJ_{\delta\delta}$ or DJ_{dd}:

$$DJ_{\delta\delta} := \mu_R - \mu_L \tag{2.48}$$

while $Q(\sigma_R + \sigma_L)$ takes into account the random jitter component and is indicated as usual by RJ. Note that the value of RJ depends on the chosen Q factor, thus on the chosen level of probability of jitter exceeding the peak bounds. In the context of jitter decomposition, the peak-peak value $\rho\rho$ is often referred to as *Total Jitter* and indicated with TJ. From Equation 2.44 the relationship between TJ, DJ, and RJ can be written as:

$$TJ = DJ_{\delta\delta} + RJ(Q) \tag{2.49}$$

where we emphasize that RJ is a function of the factor Q.

It is important to underline that a value of TJ without the specification of the probability level assumed, or of the Q-factor used, is meaningless. The specification of TJ must always be accompanied by the probability value for which that TJ has been calculated.

EXAMPLE 1   In the case shown in Figure 2.15, the $DJ_{\delta\delta}$ is equal to 1.9ps $- (-3.4\text{ps}) =$ 5.3ps. The RJ part, taking factor $Q = 7$, is equal to $7\cdot(1+1)\text{ps} = 14\text{ps}$. The peak positive $\rho^+ = 1.9 + 7 = 8.9\text{ps}$ and the peak negative $\rho^- = -3.4 - 7 = -10.4\text{ps}$. The corresponding TJ is thus 19.3ps and is associated to a probability of $(0.19+0.45)\cdot 1.28\cdot 10^{-12}$ $= 0.82\cdot 10^{-12}$ of finding a jitter sample outside the peaks. Note that if $Q = 3$, then RJ $= 6\text{ps}$, $\rho^+=4.9\text{ps}$, $\rho^- = -6.4\text{ps}$ and TJ = 11.3ps, which is much lower than the previous value, but, in this case, the probability associated with this numbers would be only $0.86\cdot 10^{-3}$.

The reader may ask why the concept of dual Dirac DJ ($DJ_{\delta\delta}$) must be introduced in addition to the "regular" DJ. The fact is that these two quantities rarely coincide. The $DJ_{\delta\delta}$ is an abstraction whose purpose is to *model* correctly the TJ. It is derived purely mathematically from the tail-fitting algorithm as the distance between the center of the two fitted Gaussian curves, and does not represent a real jitter component present in the system. As shown in Figure 2.14, due to the convolution process of RJ with DJ, the roll-off points of the resulting jitter are moved inwards with respect to the DJ peak values. As a result, the available algorithms commonly used to extract the DJ tend to move the centers of the Gaussian curves fitting the tails inwards, underestimating the DJ. In this sense, the algorithms optimizing also the amplitude A are much more accurate than those assuming $A = 1$, but still tend to underestimate the DJ component. Therefore $DJ_{\delta\delta}$ is always less than or equal to DJ, and equality is reached mostly only when the real DJ jitter component is a perfect Dual Dirac. Using DJ instead of $DJ_{\delta\delta}$ in Formula 2.49 yields an incorrect TJ value for the desired probability level.

A final note on the tail-fitting approach is in order. Although the concept is straightforward, its practical implementation faces severe challenges. The main difficulty lies in the fact that the tail region of the PDF to be fitted must be far out enough so as not to be significantly perturbed by DJ components, which could cause erroneous estimations of the RJ parameters. These extreme parts of the histogram tails are also the parts with the lowest number of hits. Variability on the tails due to the limited number of hits, as well as outliers due to perturbed measurements, create very difficult conditions for a reliable and robust fit. Several algorithms have been developed in the last decade to enhance the accuracy and reliability of tail fitting and their detailed descriptions go beyond the scope of the book. Section 10.4 briefly outlines the basic method used in the most popular approaches, involving the concept of *Q-scale* and *normalized Q-scale*. The interested reader is invited to see [15–23] for further information.

3 Jitter and Phase Noise

This chapter recalls the definitions and the insights introduced in Chapter 2 and elaborates on them in mathematical terms. It gives the reader a more complete understanding of the subject from a mathematical point of view, and provide tools and techniques to analyze quantitatively jitter issues in real systems. The mathematical foundations are reviewed in Appendix A.

In the main part of the chapter, the relation between phase noise and jitter is treated in depth. Examples of phase noise profiles are taken from those most common in practice, and their corresponding jitter is derived. The effect of spectral spurious tone on jitter is also considered and analyzed.

3.1 Basic Relationship Between Jitter and Excess Phase

In the following sections, the jitter concepts presented in the previous chapter will be related to the excess phase. Unless otherwise noted, we will assume a clock signal as described by Equation 2.2, and we refer the reader to Section 2.1 for a more generic discussion of the concept of an ideal clock and excess phase.

3.1.1 Excess Phase and Absolute Jitter in the Time Domain

The theory of phase noise is based on the assumption that the excess phase due to noise in electronic systems is a continuous-time process. This assumption is very well justified in practice. Indeed, even though the ultimate sources of noise – the charge carriers in the devices – are intrinsically quantized, noise events normally involve such a large number of particles and occur at such a high frequency that, for all practical purposes, even on a very small time scale, their effects can be considered continuous.

Jitter, on the other hand, can be defined and measured only when a clock transition is occurring, so it is intrinsically a discrete-time process. In this section we clarify the dependencies of the two processes.

If we assume a generic clock, represented by $x(\omega_0 t)$ periodic in $T_0 = 2\pi/\omega_0$, where the term $\omega_0 t$ is the total phase of the signal, the time instants of the clock transitions can be defined as the time when the signal crosses a given constant threshold value with, e.g., positive slope. By properly choosing the origin of the time axis, the k-th clock transition occurs when the total phase is equal to $2\pi k$. In the ideal, noiseless case, these

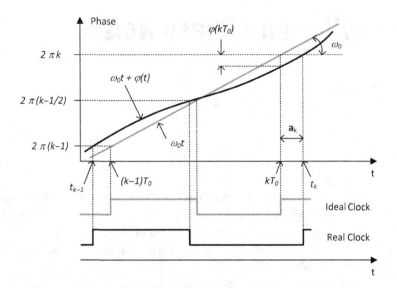

Figure 3.1 Excess phase versus time. At the bottom, an ideal clock signal and a jittered one are illustrated. The corresponding total phases are shown in the graph at the top. The voltage signals toggle as the total phase crosses multiples of π.

instants are such that $\omega_0 t = 2\pi k$, so that $t = kT_0$. In presence of noise the total phase of the signal can be modeled as $\omega_0 t + \varphi(t)$, where $\varphi(t)$ is the excess phase. Under these conditions, the real clock transitions will occur at the instants t_k such that:

$$\omega_0 t_k + \varphi(t_k) = 2\pi k \qquad (3.1)$$

as shown in Figure 3.1. Since the absolute jitter $\mathbf{a}_k$ is defined as $t_k - kT_0$ (see Equation 2.4), dividing both sides of the equation by ω_0 and rearranging the terms, $\mathbf{a}_k$ can be expressed as:

$$\mathbf{a}_k = -\frac{\varphi(t_k)}{\omega_0}. \qquad (3.2)$$

The value of $\varphi(t_k)$ can be calculated by expanding the function $\varphi(t)$ in Taylor's series around kT_0. Replacing the distance $t_k - kT_0$ by $\mathbf{a}_k$, we obtain:

$$\varphi(t_k) = \varphi(kT_0) + \mathbf{a}_k \dot{\varphi}(kT_0) + \frac{\mathbf{a}_k^2}{2}\ddot{\varphi}(kT_0) + \cdots \qquad (3.3)$$

Assuming that the terms of order two or above are negligible compared to the first-order term, and combining the last two equations, the absolute jitter can be expressed as:

$$\mathbf{a}_k = \frac{-\varphi(kT_0)}{\omega_0 + \dot{\varphi}(kT_0)}. \qquad (3.4)$$

Finally, if the excess phase is sufficiently well behaved, meaning that its change rate is much smaller than the change rate of the unperturbed phase ($\dot{\varphi}(kT_0) \ll \omega_0$), the absolute jitter can be expressed as:

$$\mathbf{a}_k = \frac{-\varphi(kT_0)}{\omega_0}. \qquad (3.5)$$

From the expression above, the absolute jitter is the discrete-time random process obtained by sampling the continuous-time excess phase random process $\varphi(t)$ at equidistant instants T_0 and scaling it by the radian frequency ω_0.

3.1.2 Excess Phase and Absolute Jitter in the Frequency Domain

The previous section addressed the relationship between absolute jitter and excess phase in the time domain. In this section we will look at the same topic but from a frequency domain perspective.

An important concept in the analysis of random processes in the frequency domain is the Power Spectral Density (PSD). Here we will summarize the main concept of the PSD. A thorough discussion of the PSD can be found in many textbooks (see, e.g., [1]) and a very brief summary is presented in Appendix A.

The PSD is, for random processes, what the spectrum is for deterministic signals. It tells us how the power of the random process is distributed over the frequency axis. A flat PSD means that all frequency components are equally present in the process. This is the case for the thermal noise produced by a resistor, for instance, which turns out to be flat up to very high frequencies. If a process has a flat PSD it is also called *white*, since all frequencies (colors) are contributing equally to the total process. Passing a white noise through a low-pass filter, e.g., by connecting a capacitor in parallel to the resistor, dampens the high frequency components of the noise. In this case, the PSD turns out to be still flat below the bandwidth of the filter, and rolls off at higher frequencies. By integrating the PSD between two frequencies, we obtain the amount of power contributed to the process by those frequencies. Integrating the PSD over the whole frequency axis gives the total power of the process, which is equal to its variance. From a mathematical standpoint, the PSD is obtained as the Fourier transform of the autocorrelation of the process; it is an even function of frequency and can assume only positive real values. As an example, a flat PSD implies that the autocorrelation is a Dirac function, meaning that the samples of the process are completely uncorrelated, no matter how close in time they are. In what follows we will apply the concept of PSD to find a relation between absolute jitter and excess phase in the frequency domain.

As explained in Section 3.1.1, the jitter is a sampled and scaled version of the excess phase. From the basic theory of random processes, it follows that the autocorrelation of jitter R_a is obtained by sampling the autocorrelation of the excess phase R_φ:

$$R_a(kT_0) = \frac{R_\varphi(kT_0)}{\omega_0^2} \tag{3.6}$$

and, finally, the PSD of the jitter can be obtained by folding and scaling the PSD of the excess phase:

$$S_a(f) = \frac{1}{\omega_0^2} \sum_{n=-\infty}^{+\infty} S_\varphi(f + nf_0) \tag{3.7}$$

and is defined in the frequency range $-f_0/2 < f < +f_0/2$ (see Figure 3.2).

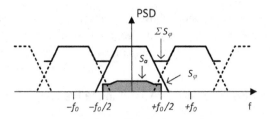

Figure 3.2 Relationship between PSD of the excess phase and PSD of jitter.

Note that if $S_\varphi(f)$ is zero outside the interval $[-f_0/2, +f_0/2]$, there is no superposition of the replicas. This is often the case in many applications, where the power of the excess phase due to thermal or flicker noise decreases with increasing frequencies, due to intrinsic low-pass or integration effects. In such cases the residual power at frequency higher than $f_0/2$ can be neglected. For these cases, Equation 3.7 simplifies to:

$$ S_\mathbf{a}(f) = \begin{cases} \dfrac{S_\varphi(f)}{\omega_0^2} & \text{for} & \dfrac{-f_0}{2} \le f \le \dfrac{+f_0}{2} \\ 0 & \text{otherwise} \end{cases} \qquad (3.8) $$

so that the PSD of jitter is identical to the PSD of the excess phase, apart from a scaling factor.

On the other hand, if excess phase at frequencies larger than $f_0/2$ is present, it will manifest itself as jitter at a frequency between 0 and $f_0/2$. This phenomenon is equivalent to the aliasing of analog signals in a sampling process in which the Nyquist criterion is not satisfied. In the case of jitter, this makes physical sense, since noise components separated in frequency by multiples of f_0 cannot be distinguished when observing the signal at time multiples of $T_0 = 1/f_0$. Figure 3.3 illustrates this concept with an example. It can be seen, that, since jitter can be measured only at the clock rising edges, the noise component at $5f_0/8$ has exactly the same effect on the signal as a component at $3f_0/8$. Looking at Figure 3.3, it might appear that the two noise components superimpose linearly, so that the total noise power at $3f_0/8$ should be four times higher that the noise of a single component, assuming both have the same power. However, we are dealing here with random noise processes, so that the position of the second component on the time axis is randomly distributed with respect to the position of the first. On average, therefore, the jitter PSD at frequency $3f_0/8$ is just the sum of the two noise powers, as correctly described by the folding function. The proof is left as an exercise for the reader.

3.1.3   Voltage to Excess Phase Transformations: Random Noise

In the previous section, it was shown that the PSD of the jitter can be derived from the PSD of the excess phase. The excess phase, however, is not an easily measurable quantity like a voltage or a current. The question then is how to obtain the PSD of the excess phase from measurements of the clock signal, which is normally a voltage signal.

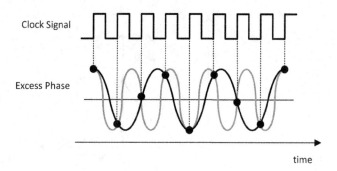

Figure 3.3 Excess phase aliasing. The clock signal is shown as a rectangular signal, and two excess phase sinusoidal components at frequencies $5f_0/8$ (dashed) and $3f_0/8$ (solid) are shown superimposed.

Let us represent the voltage signal as a sinusoid having a nominal radian frequency $\omega_0 = 2\pi f_0 = 2\pi/T_0$:

$$v(t) = A(t)\sin(\omega_0 t + \varphi(t)). \tag{3.9}$$

We are only interested in the excess phase, therefore we will assume that the amplitude $A(t)$ is constant.

This signal can be also written as:

$$v(t) = A\left[\sin(\omega_0 t)\cos\varphi(t) + \cos(\omega_0 t)\sin\varphi(t)\right] \tag{3.10}$$

and, under the assumption that $\varphi(t) \ll 1$ (also called *narrow angle* assumption), as:

$$v(t) \approx A\sin(\omega_0 t) + A\varphi(t)\cos(\omega_0 t). \tag{3.11}$$

The phase modulated sinusoid has been decomposed into the sum of an undisturbed carrier plus additive noise modulated by a quadrature carrier, so that the additive noise is at its maximum where the carrier is near zero. It is left as an exercise for the reader to prove that the time deviation from the nominal crossing point is consistent with Equation 3.5.

Based on this approximation, the autocorrelation of the voltage signal $v(t)$ can be calculated as:

$$R_v(t,\tau) = \frac{A^2}{2}\big\{\cos(\omega_0\tau) - \cos(2\omega_0 t + \omega_0\tau) + [\cos(2\omega_0 t + \omega_0\tau)$$
$$+ \cos(\omega_0\tau)]\,R_\varphi(\tau)\big\} \tag{3.12}$$

$R_v(t,\tau)$ is periodic in t with a period of π/ω_0 (*cyclostationary process*) and its average value over one period is:

$$\bar{R}_v(\tau) = \frac{\omega_0}{\pi}\int_0^{\pi/\omega_0} R_v(t,\tau)dt = \frac{A^2}{2}\cos(\omega_0\tau)\left[1 + R_\varphi(\tau)\right]. \tag{3.13}$$

Applying a Fourier transformation, the two-sided PSD can be calculated as:

$$S_v(f) = \frac{A^2}{4}\left[\delta(f - f_0) + \delta(f + f_0) + S_\varphi(f - f_0) + S_\varphi(f + f_0)\right] \tag{3.14}$$

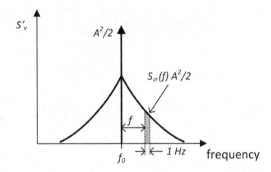

Figure 3.4 Relation between voltage spectrum and PSD of excess phase.

thus the one-sided PSD, defined for $f \geq 0$, is:

$$S_v'(f) = \frac{A^2}{2}\left[\delta(f - f_0) + S_\varphi(f - f_0)\right] \tag{3.15}$$

as shown in Figure 3.4.[1] This equation reveals that, due to the modulation process described by Equation 3.9, the baseband spectrum of the excess phase at a given frequency f is up-converted to the frequency $f_0 + f$ and appears then as sideband around the oscillation frequency f_0. By reversing these equations and denoting now with f the frequency offset from the carrier, the two-sided PSD of the excess phase can be derived (for $f \neq 0$) from the one-sided or two-sided voltage PSD as:

$$S_\varphi(f) = \frac{S_v(f_0 + f)}{A^2/4} = \frac{S_v'(f_0 + f)}{A^2/2}. \tag{3.16}$$

In other words, the PSD of the excess phase at a given frequency f can be obtained as the power of the voltage signal within 1 Hz band at an offset f from the carrier divided by the power of the carrier itself.

The considerations above assume a pure sinusoidal signal. In practical applications, however, the clock signal is not a pure sinusoid and its spectrum contains harmonics at multiple frequencies of the fundamental one. If we assume a generic periodic signal $x(t)$, choose the time origin so that $x(0) = 0$, and, for the sake of simplicity, assume that the signal is odd-symmetrical about the origin (meaning $x(-t) = -x(t)$), it can be expanded in Fourier series as:

$$x(t) = \sum_{n=1}^{+\infty} c_n \sin(n\omega_0 t). \tag{3.17}$$

If the signal is affected by absolute jitter a_k changing the instants of the zero crossings but not the shape of the signal itself, the term t in the expression above can be replaced by $t - a(t)$ (note that the minus sign is introduced for consistency with the definitions of absolute jitter and excess phase):

$$x(t - a(t)) = \sum_{n=1}^{+\infty} c_n \sin(n\omega_0 t - n\omega_0 a(t)). \tag{3.18}$$

[1] For definitions of one-sided and two-sided PSD, refer to Section A.2.8.

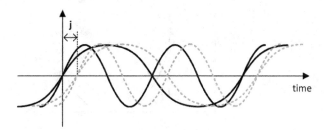

Figure 3.5 Rigid shift of the harmonic components of a period signal affected by jitter j.

The auxiliary continuous time function $\mathbf{a}(t)$ is introduced as an extension of the concept of absolute jitter in the continuous time domain, and is equal to $\mathbf{a}_k$ at the zero crossings. This function has no real physical meaning, but is helpful in deriving the considerations in the present section.

It can be recognized that $-n\omega_0\mathbf{a}(t)$ is the excess phase φ_n affecting the n-th harmonic. Since $-\omega_0\mathbf{a}(t)$ is the excess phase affecting the fundamental frequency, indicated by φ in the case of the pure sinusoid above, it follows that:

$$\varphi_n = n\varphi \tag{3.19}$$

the excess phase of the n-th harmonic is n times larger than the excess phase of the fundamental.

This can be understood intuitively considering that the excess phase is a measure of jitter relative to the period of the signal. Since the jitter affects the zero crossing point but not the shape of the signal, the waveform is rigidly shifted on the time axis, so that all frequency components are shifted by the same amount regardless of their periods (see Figure 3.5). The amount of shift affecting a frequency component with a period n times smaller must therefore lead to an excess phase n times larger.

For the PSD, this means:

$$S_{\varphi_n}(f) = n^2 S_\varphi(f). \tag{3.20}$$

This expression shows that the PSD of the excess phase around the n-th harmonic in n^2 times larger than the one around the fundamental. Following the same approach as in the case of a pure sinusoid, $S_{\varphi_n}(f)$ can be derived from measurements as the power in a 1Hz bandwidth at an offset f from the n-th harmonic divided by the energy of the n-th harmonic. However, this result has to be further divided by n^2 to obtain $S_\varphi(f)$. Figure 3.6 shows graphically the relationship between the PSD of the excess phase and the voltage spectrum of a non-sinusoidal periodic signal.

3.1.4 Voltage to Excess Phase Transformations: Modulation

In the previous section, the excess phase $\varphi(t)$ has been considered to be a random process having a given PSD $S_\varphi(f)$. It is also important to consider cases when the excess phase is a deterministic periodic signal, since such situations are often encountered in practice, for instance in a clock distribution network with periodic disturbance on the supply, or in a PLL, due to the periodic nature of the phase comparison and VCO voltage adjustments.

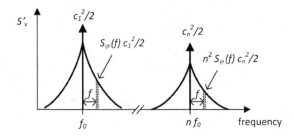

Figure 3.6 Relationship between PSD of excess phase and spectrum of signal with harmonics.

Let us assume then that the excess phase is a sinusoidal signal with a frequency $\omega_m \ll \omega_0$ and amplitude Φ_m, $\varphi(t) = \Phi_m \sin(\omega_m t)$. The voltage signal in Equation 3.9 is then:

$$v(t) = A \sin[\omega_0 t + \Phi_m \sin(\omega_m t)] \tag{3.21}$$

where the amplitude $A(t)$ is again assumed to be time-independent. For a small angle modulation $\Phi_m \ll 1$, this can be expanded in series by use of Bessel functions of the first kind J_n as:

$$v(t) = A J_0(\Phi_m) \sin(\omega_0 t) + A J_1(\Phi_m) [\sin(\omega_0 t + \omega_m t) - \sin(\omega_0 t - \omega_m t)]$$
$$+ A J_2(\Phi_m) [\sin(\omega_0 t + 2\omega_m t) - \sin(\omega_0 t - 2\omega_m t)] + \cdots \tag{3.22}$$

It can be seen that the modulation of the excess phase at frequency ω_m translates into discrete tones at offset multiples of ω_m around the carrier ω_0. The level of the sideband tones with respect to the carrier can be calculated easily. Using the small angle assumption, the first three Bessel functions can be approximated by $J_0(\Phi_m) \approx 1$, $J_1(\Phi_m) \approx \Phi_m/2$ and $J_2(\Phi_m) \approx \Phi_m^2/8$. For small Φ_m, the tones at frequencies $2\omega_m$ or higher are much weaker than those at ω_m and can normally be ignored. Applying these expressions to Equation 3.22, the difference in power level between the first tone and the carrier is called Spurious to Carrier Ratio (SCR) and typically expressed in dBc (dB from carrier):

$$SCR_{dBc} = 10 \log \left(\frac{\Phi_m}{2} \right)^2 \text{ dB}. \tag{3.23}$$

Conversely, from spectrum measurements, the excess phase modulation amplitude can be derived from the SCR_{dBc} as:

$$\Phi_m = 2 \cdot 10^{(SCR_{dBc}/20)}. \tag{3.24}$$

Figure 3.7 summarizes the voltage to excess phase transformations for the case of random noise (Equation 3.16) as well as for a deterministic modulation (Equations 3.23 and 3.24).

EXAMPLE 2   It is useful to get a sense of the order of magnitude of the SCR typically found in modern clock generation circuits and the corresponding excess modulation amplitude. An SCR of −60dB represents an excellent value, normally obtainable only with ad hoc techniques, aimed at reducing the deterministic periodical disturbances.

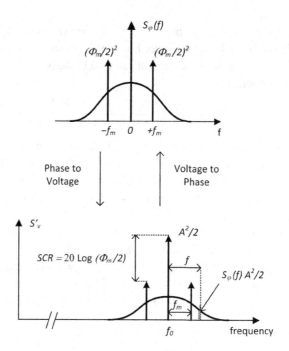

Figure 3.7 Overview of voltage to excess phase transformations.

Using the equation above, it can be seen that this value corresponds to an excess phase modulation of $\Phi_m = 2\,\text{mrad}$. Every 6dB added to the SCR multiplies the modulation amplitude by two.

3.1.5   Definition of Phase Noise

Although the concept of phase noise is broadly used, its definition was subject to discussion in recent years. In this book, phase noise will be indicated with the symbol $\mathcal{L}$, in accordance with the common usage in the literature. Given a voltage periodic signal with frequency f_0, power P, one-sided *voltage* PSD S'_v and affected by random excess phase modulation φ, the phase noise is traditionally defined as the ratio of the power of the signal in 1Hz bandwidth at offset f from the carrier, divided by the power of the carrier:

$$\mathcal{L}(f) := \frac{S'_v(f_0 + f) \text{ in 1Hz bandwidth}}{P}. \tag{3.25}$$

Note that $\mathcal{L}(f)$ is defined over positive frequencies only ($f \geq 0$). In 1999 the IEEE released the Standard 1139, revised in 2008 [24], in which the phase noise is defined as the two-sided PSD S_φ, or half of the one-sided PSD S'_φ, of the excess phase φ:

$$\mathcal{L}(f) := S_\varphi(f) = \frac{S'_\varphi(f)}{2}. \tag{3.26}$$

This standard seems not to have been widely adopted by industry or academia, but we report it for completeness. It was proven in Section 3.1.3 that, using the narrow angle assumption, the two definitions above are equivalent (see Equation 3.16). If the narrow angle condition is not satisfied, however, the two definitions differ. Following Section 3.1.3, the phase noise of a signal can also be derived from the voltage spectrum of the *n*-th harmonic as:

$$\mathcal{L}(f) := \frac{1}{n^2} \frac{S_v'(nf_0 + f) \text{ in 1Hz bandwidth}}{\text{power of the } n\text{-th harmonic}}. \tag{3.27}$$

Note that in this section we referred to *one-sided* and *two-sided* PSD, and we avoided the use of the terms *single sideband* and *double sideband*. The use of these terms is often a source of misunderstanding, so they should be treated with caution. A discussion of the different engineering definitions of PSD and their usage in this book is reported in Appendix A.2.8.

Also note that in this section, as well as in the following ones, the symbol f denotes the frequency offset from the carrier, and not the absolute frequency. In general, the argument of the phase noise $\mathcal{L}(\cdot)$ represents always the frequency offset from the carrier, independent of the symbols used for it. Some authors use the symbol f_m for it, but for simplicity we keep f, unless there is risk of misunderstanding.

3.2   From Phase Noise to Jitter

The following sections will illustrate how to compute absolute, period and N-period jitter starting from the phase noise profile.

3.2.1   Absolute Jitter

Using the Wiener–Khinchin theorem (see Section A.2.7) it is possible to easily derive the variance of the absolute jitter via integration of the corresponding PSD:

$$\sigma_a^2 = \int_{-f_0/2}^{+f_0/2} S_a(f)\, df. \tag{3.28}$$

It has been shown in Section 3.1.2 that $S_a(f)$ is a scaled and folded version of the PSD $S_\varphi(f)$, so that the area of S_a from $-f_0/2$ to $+f_0/2$ is equal to the area of $S_\varphi(f)$ over the frequency range from $-\infty$ to $+\infty$, scaled by $1/\omega_0^2$. Since, using the narrow angle assumption, $S_\varphi(f)$ is equal to the phase noise $\mathcal{L}(f)$, the RMS absolute jitter can then be calculated as:

$$\sigma_a = \sqrt{\frac{2}{\omega_0^2} \int_0^{+\infty} \mathcal{L}(f)\, df} \tag{3.29}$$

where the integral is carried out only over positive frequencies and multiplied by two, considering that phase noise is normally symmetrical about the zero frequency. It must be remembered that, in the expression above, and in all expressions that follow, the numerical value of the phase noise in the integral must be expressed in linear units, not in logarithmic ones ($\mathcal{L}[linear] = 10^{(\mathcal{L}[dBc/Hz]/10)}$).

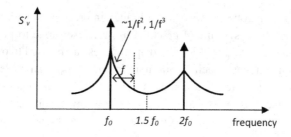

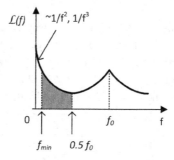

Figure 3.8 Voltage spectrum of a noisy periodic signal with multiple harmonics (top) and corresponding phase noise plot (bottom). The upper noise sideband of the carrier superimposes with the lower sideband of the second harmonic.

There are two inherent problems with the expression above, both related to the integration limits. The first one concerns the integration towards $+\infty$. As shown in Section 3.1.3, the phase noise is calculated as the ratio of sideband noise power over carrier power. In most of the practical cases, the signal under investigation is not a perfect sinusoid, so the spectrum presents multiple harmonics. The upper noise sideband of the first harmonic at f_0 will then overlap with the lower noise sideband of the second harmonic at $2f_0$, as illustrated in Figure 3.8.

The degree of superposition depends on the shape of the signal, but we can expect that, starting from frequencies above $1.5f_0$, the noise of the second harmonic will start to dominate over the noise of the first. We have shown in Figure 3.6 that the phase noise around the various waveform harmonics is the scaled replica of the phase noise around the fundamental, and that each of these replicas alone account for the total jitter. For this reason, to compute the jitter correctly we can focus solely on the phase noise around one of these harmonic and thus limit the upper integral of Equation 3.29 to an offset of about $f_0/2$ from the carrier f_0.

The second concern is related to the lower integration limit equal to zero. All physically implementable oscillators and signal sources show a $1/f^2$ or even $1/f^3$ behavior of the phase noise versus frequency f for offsets very close to the carrier (see Figure 3.8). It is well known from elementary analysis that the integrals of such functions down to zero diverge to infinity. This result is indeed in accordance with physics: if the oscillator output is observed for an infinite time, the absolute jitter will increase indefinitely.

Fortunately, observation time is limited in practical situations and, additionally, almost all electronic systems of practical use are insensitive to jitter fluctuations at very low frequencies, so that at least two approaches can be used to overcome this issue.

In the first approach, a minimum frequency f_{min} can be defined based on the analysis of how jitter affects the electronic system under investigation or the maximum observation time. It is probably the most common way of solving the problem, although not the most correct one. In this case Equation 3.29 can be written as:

$$\sigma_{\mathbf{a}} = \sqrt{\frac{2}{\omega_0^2} \int_{f_{min}}^{+f_0/2} \mathcal{L}(f)\, df}. \tag{3.30}$$

The alternative, and more exact, way is to analyze how jitter at different frequencies affects the performance of the system. This can be expressed in terms of a jitter transfer function $H_{sys}(f)$ in the frequency domain, which weights the jitter components according to their impact. The RMS value of the absolute jitter can then be calculated by integrating the phase noise filtered by this transfer function:

$$\sigma_{\mathbf{a}} = \sqrt{\frac{2}{\omega_0^2} \int_0^{f_0/2} \mathcal{L}(f)\, |H_{sys}(f)|^2\, df}. \tag{3.31}$$

Typically $H_{sys}(f)$ shows a high-pass-like behavior, so that the integral in the expression above will not diverge for integration down to zero. Figure 3.9 illustrates this concept in the case of a phase noise profile growing as $1/f^2$ for low frequencies and $H_{sys}(f)$ having a first-order high-pass characteristic. In this case, the filtered phase noise exhibits a flat profile for low frequencies, and can be integrated without problems down to very low frequency offsets. An example of the application of this method will be given in Section 7.2.3.

We will now apply the previous expressions to four of the most common phase noise profiles: flat, $1/f^2$, PLL-like, and $1/f^3$.

Flat Phase Noise Profile

In this section, the RMS absolute jitter corresponding to a flat phase noise is derived (see Figure 3.10). Assuming:

$$\mathcal{L}(f) = \begin{cases} \mathcal{L}_0 & \text{if } 0 < f < f_{max} \\ 0 & \text{otherwise} \end{cases} \tag{3.32}$$

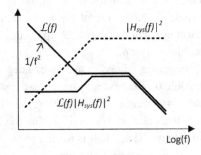

Figure 3.9 Filtering of the phase noise before integration to obtain the RMS absolute jitter.

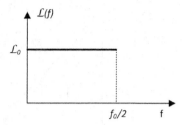

Figure 3.10 Flat phase noise profile.

and substituting this expression into Equation 3.29, the absolute jitter can be found using:

$$\sigma_{\mathbf{a}} = \frac{1}{\omega_0}\sqrt{2\mathcal{L}_0 f_{max}}.$$ (3.33)

If the white phase noise band extends to $f_{max} = f_0/2$, then, due to the repetition of the spectrum in the frequency domain, the PSD of the jitter random process j is flat for $-\infty < f < +\infty$. In such a case, a very simple expression for the ratio of the RMS absolute jitter to the clock period is found:

$$\frac{\sigma_{\mathbf{a}}}{T_0} = \frac{\sqrt{\mathcal{L}_0 f_0}}{2\pi}.$$ (3.34)

EXAMPLE 3   For $f_0 = 1\text{GHz}$ and $\mathcal{L}_0 = 10^{-13}$ (corresponding to a value in dB of -130dBc/Hz), the RMS absolute jitter is equal to 0.16% of the clock period, namely 1.6 ps.

$1/f^2$ Phase Noise Profile

Neglecting the presence of flicker noise and of the flat region for very high frequencies, to a first approximation the phase noise profile of a free-running oscillator can be described as a constant 20dB/dec frequency roll-off (see [25], [26]), as shown in Figure 3.11:

$$\mathcal{L}(f) = \frac{\mathcal{L}_1 f_1^2}{f^2}.$$ (3.35)

As explained before, this profile cannot be integrated down to zero, so that Equation 3.30 must be used, yielding the absolute jitter normalized to the period:

$$\frac{\sigma_{\mathbf{a}}}{T_0} = \sqrt{\frac{\mathcal{L}_1 f_1^2}{2\pi^2}\left(\frac{1}{f_{min}} - \frac{1}{f_{max}}\right)}$$ (3.36)

where the upper integration limit in Equation 3.30 has been replaced by f_{max} for more flexibility. Note that f_{min} is related to the observation time, as stated above. The longer we observe the device under test, the smaller f_{min} must be. Note that the final value mainly depends on f_{min}, if f_{max} is at least 10 times larger than f_{min}.

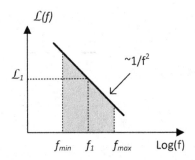

Figure 3.11 Simplified phase noise profile of a free-running oscillator.

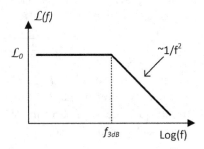

Figure 3.12 Simplified phase noise profile of a PLL.

Simplified PLL Phase Noise Profile

To a first approximation, the phase noise profile of a PLL can be described as a flat region at level $\mathcal{L}_0$ at low frequencies and a 20dB/dec roll-off for frequencies above the loop bandwidth f_{3dB}, as shown in Figure 3.12:

$$\mathcal{L}(f) = \frac{\mathcal{L}_0}{1 + (f/f_{3dB})^2}. \tag{3.37}$$

In PLL design, the flat $\mathcal{L}_0$ region up to f_{3dB} is called *in-band* phase noise and is mainly due to noise on the reference clock, in circuits in the forward path before the loop-filter, and in the feedback path. The $1/f^2$ part is a combination of all possible noise sources of the PLL, including loop filter and VCO. Typically, however, a high-performance PLL is designed so that the phase noise in this region is limited by the VCO noise.

The expression above can be easily integrated over the whole frequency range. Applying Equation 3.29 and remembering that $\int 1/(1 + a^2)da = \arctan(a)$, the RMS value relative to the clock period can be written as:

$$\frac{\sigma_a}{T_0} = \sqrt{\frac{\mathcal{L}_0 f_{3dB}}{4\pi}} \tag{3.38}$$

which is a fairly simple expression to remember. The same result can be obtained by remembering that the area below a profile like the one shown in Figure 3.12 can be calculated by multiplying $\mathcal{L}_0$ by $f_{3dB}\pi/2$, where $f_{3dB}\pi/2$ is the so-called *equivalent noise bandwidth*. It is left as an exercise for the reader to prove this equivalence.

EXAMPLE 4   In paper [27], the authors present a high-performance digital PLL with programmable bandwidth. The phase noise profile is measured at an output frequency of 3.61GHz for three different bandwidth values: 300kHz, 1MHz and 2MHz. The in-band phase noise for the three cases is -100dBc/Hz, -102dBc/Hz and -104dBc/Hz respectively. The phase noise of this design is particularly smooth and doesn't show significant peaking, so that it is easy to calculate the corresponding absolute jitter with great accuracy by applying Equation 3.38. The result is 428fs, 621fs, and 697fs for the three cases, respectively: very close to the published numbers (440fs, 660fs, and 680fs respectively).

EXAMPLE 5   Paper [28] presents a 3GHz digital PLL with bandwidth of 1.8MHz and an in-band phase noise of -102.5dBc/Hz. In this case, though, the phase noise profile is not completely flat, showing a peaking of about 3dB around 1.2MHz frequency offset. By applying Expression 3.38 with $\mathcal{L}_0 = -102.5$, the RMS absolute jitter turns out to be 0.95ps. If $\mathcal{L}_0 = -99.5$ is used instead, the result is 1.34ps. One might think that by numerically integrating the exact phase noise profile, the result should be between these two numbers. In reality, and maybe a bit surprisingly, the result is 0.86ps. The reason for this difference is that the out of band phase noise for this PLL is dominated not by the VCO, but by other blocks. The noise produced by those blocks is filtered by a low pass transfer function of second order due to the particular design of the loop filter, so that the out of band phase noise profile decreases with -40dB/dec, instead of the -20dB/dec assumed in deriving Equation 3.38.

It is instructive to investigate which parts of the phase noise profile shown in Figure 3.12 contribute the most to the resulting absolute jitter. To do so, the absolute jitter is calculated by integrating the phase noise from 0 to a moving upper limit f_{max}. The result is:

$$\frac{\sigma_a(f_{max})}{T_0} = \sqrt{\frac{\mathcal{L}_0 f_{3dB}}{2\pi^2} \arctan\left(\frac{f_{max}}{f_{3dB}}\right)}. \tag{3.39}$$

Except for a constant multiplicative factor, the behavior of this function is described by the function:

$$\sqrt{\frac{2}{\pi} \arctan\left(\frac{f_{max}}{f_{3dB}}\right)}. \tag{3.40}$$

Figure 3.13 shows a plot of this function versus $\log(f_{max}/f_{3dB})$. It is interesting to note that the maximum slope is reached around the bandwidth of the PLL f_{3dB}. Therefore, when looking at the phase noise profile on a logarithmic frequency axis, most of the absolute jitter is coming from frequencies around the PLL bandwidth. From Figure 3.13 it can be seen that 60% of the absolute jitter comes from frequencies between one tenth and two times the PLL bandwidth. This point deserves to be highlighted further. Phase noise profiles are usually plotted on logarithmic frequency scales, so quick glances at them can be deceiving. One would naturally think that frequency intervals of similar "visual" length on the x-axis would yield the same amount of jitter, and that all parts of the in-band phase noise spectrum are equally important. But this ignores the nature

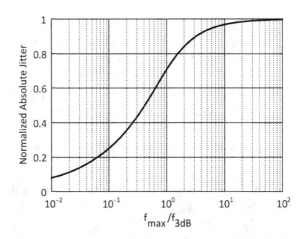

Figure 3.13 Normalized absolute jitter for a simplified PLL phase noise profile, as function of the upper integration limit f_{max} normalized to the PLL bandwidth f_{3dB}.

of logarithmic scales: a frequency interval of a given visual length around a certain frequency f is actually *ten times* larger that an interval of the same visual length around a frequency a decade further left on the x-axis, $f/10$. Therefore each logarithmic unit to the right contributes ten times more jitter than the one before it.[2] As a consequence, most of the effort in PLL design to reduce absolute jitter should be devoted to controlling the phase noise around the PLL bandwidth. Any noise contributors around this area, as well as any peaking in the phase noise profile, are particularly harmful.

$1/f^3$ Phase Noise Profile

Here we derive the absolute jitter corresponding to a $1/f^3$ characteristic of the phase noise. This phase noise profile is typical for the low frequency behavior of a free-running oscillator with internal flicker noise sources. Assuming

$$\mathcal{L}(f) = \frac{\mathcal{L}_1 f_1^3}{f^3} \tag{3.41}$$

and integrating this function down to zero would lead to non-convergence of the result, consistent with the physical reality, in which flicker noise processes have infinite variance. However, in practical applications, the process is observed only for a limited amount of time, and we can define a lower bound frequency f_{min} for the integration of the process. Substituting Expression 3.41 into Equation 3.29 and limiting the integral from f_{min} to f_{max} we obtain for the ratio of the RMS absolute jitter to the clock period:

$$\frac{\sigma_{\mathbf{a}}}{T_0} = \sqrt{\frac{\mathcal{L}_1 f_1^3}{4\pi^2}\left(\frac{1}{f_{min}^2} - \frac{1}{f_{max}^2}\right)}. \tag{3.42}$$

---

[2] The same is of course true for classic noise; looking at noise on a logarithmic scale is misleading. To fully appreciate its frequency distribution, noise must be viewed on a linear scale.

It is interesting to note that if f_{max} is even only one decade higher than f_{min}, the second term in the round parentheses contributes only 1%. So, in many practical applications, it makes sense to neglect the limit for higher frequencies and obtain a simplified expression:

$$\frac{\sigma_a}{T_0} = \frac{\sqrt{\mathcal{L}_1 f_1^3}}{2\pi f_{min}}. \tag{3.43}$$

3.2.2 N-Period and Period Jitter

In several practical applications, the N-period jitter determines the performance of the system under consideration. We have already seen that the N-period jitter on the clock at the input of a frequency divider determines the period jitter at its output, and thus, for instance, the maximum achievable speed of a digital synchronous system working on that clock. This topic will be addressed further in Chapter 5. Another example is found in line synchronization in video applications. In filling up the pixels of a display, the video clock is synchronized to a master clock only at the beginning of each line. The last pixel of each line is thus affected by N-period jitter with N equal to the number of pixels in a line.

Due to its importance, this section is dedicated to deriving the relation between phase noise and N-period jitter.

Recalling its definition as given in Equation 2.1, N-period jitter is a discrete time random process, whose value $\mathbf{p}_k(N)$ is equal to $\mathbf{a}_{k+N} - \mathbf{a}_k$.

The variance of the N-period jitter can be expressed in the following way:

$$\sigma_{\mathbf{p}(N)}^2 = E[(\mathbf{a}_{k+N} - \mathbf{a}_k)^2] = R_{\mathbf{a}}(0) - 2R_{\mathbf{a}}(N) + R_{\mathbf{a}}(0) = 2[R_{\mathbf{a}}(0) - R_{\mathbf{a}}(N)]. \tag{3.44}$$

Using Equation 3.6 we obtain, as reported in [25]:

$$\sigma_{\mathbf{p}(N)}^2 = \frac{2}{\omega_0^2}[R_\varphi(0) - R_\varphi(NT_0)]. \tag{3.45}$$

From this formula is clear that the RMS N-period jitter is completely determined by the autocorrelation function of the excess phase $\varphi(t)$.

In many practical cases, the value of the excess phase bears less and less relation to its previous values the further back in time these values lie. In other words, the excess phase process has asymptotically no memory; it is asymptotically uncorrelated. In mathematical terms this means that $\lim_{N \to +\infty} R_\varphi(NT_0) = 0$, so that:

$$\lim_{N \to +\infty} \sigma_{\mathbf{p}(N)}^2 = \frac{2}{\omega_0^2} R_\varphi(0) = \frac{2}{\omega_0^2} \cdot \text{Area}[\mathcal{L}(f)] = 2\sigma_a^2. \tag{3.46}$$

Note that this is the long-term jitter as defined in Section 2.1.5, so that $\sigma_{lt} = \sqrt{(2)}\sigma_a$. This result can be derived more directly by using Equation 2.9 and expressing the N-period jitter as the difference of two absolute jitter values N periods apart. If the excess phase is asymptotically uncorrelated, so are the two absolute jitter values on the right-hand side. Therefore, the variance of their difference for very large N is equal to the sum of their variances, leading to the result in Equation 3.46.

The RMS period jitter corresponds to the N-period jitter for $N = 1$ and can be expressed as:

$$\sigma_p^2 = \frac{2}{\omega_0^2} [R_\varphi(0) - R_\varphi(T_0)]. \tag{3.47}$$

To express the N-period jitter as a function of the phase noise $\mathcal{L}$ we use Equation 3.26 and the fact that the autocorrelation is the inverse Fourier transform of the PSD. Starting from Equation 3.45 it is easy to show that:

$$\sigma_{p(N)}^2 = \frac{2}{\omega_0^2} \int_{-\infty}^{+\infty} S_\varphi(f)(1 - e^{j2\pi fNT_0}) \, df. \tag{3.48}$$

Since the excess phase PSD is an even function of the frequency offset, we can further simplify the expression. As the integration interval is symmetrical around zero, we can remove the odd term in the complex exponential $e^{j2\pi fNT_0} = \cos(2\pi fNT_0) + j\sin(2\pi fNT_0)$. Replacing $S_\varphi(f)$ with $\mathcal{L}(f)$, finally the result is:

$$\sigma_{p(N)}^2 = \frac{4}{\omega_0^2} \int_0^{+\infty} \mathcal{L}(f)[1 - \cos(2\pi fN/f_0)] \, df$$

$$= \frac{8}{\omega_0^2} \int_0^{+\infty} \mathcal{L}(f) \sin^2(\pi fN/f_0) \, df. \tag{3.49}$$

For the cases where the integral is diverging, the integration interval can be limited from f_{min} to $f_0/2$, as explained in Equation 3.2.1. Note, however, that since $\sin^2(f)$ goes as f^2 for small f, the integration down to frequency zero of phase noise profiles of type $1/f^{-\alpha}$ with $\alpha \leq 2$ does not lead to diverging results. For these cases it is not necessary to limit the integral to f_{min}.

In the following sections this expression will be applied to the most common phase noise profiles.

Flat Phase Noise Profile

In this section we derive the RMS N-period and period jitter corresponding to a flat phase noise, as described by Equation 3.32 and repeated here for convenience:

$$\mathcal{L}(f) = \begin{cases} \mathcal{L}_0 & \text{if } 0 < f < f_{max}. \\ 0 & \text{otherwise} \end{cases} \tag{3.50}$$

Substituting this expression into Equation 3.49, the following expression for the N-period jitter can be found:

$$\sigma_{p(N)}^2 = \frac{4\mathcal{L}_0}{\omega_0^2} \left[f_{max} - \frac{\sin(2\pi f_{max}N/f_0)}{2\pi N/f_0} \right]. \tag{3.51}$$

If the white noise band extends to $f_{max} = f_0/2$, we can express the ratio of the RMS N-period jitter to the clock period as:

$$\frac{\sigma_{p(N)}}{T_0} = \frac{\sqrt{2\mathcal{L}_0 f_0}}{2\pi}. \tag{3.52}$$

The RMS value of the N-period jitter is independent of N, and is equal to the RMS period jitter and the RMS long-term jitter. This should not be surprising because the flat spectrum of the phase noise corresponds to an absolute jitter with an auto-correlation function that is a delta function. As a result, all sub-sampled sequences of the original absolute jitter sequence will be uncorrelated and, since they are stationary, will have the same statistical properties. Note also that the N-period jitter is $\sqrt{2}$ larger than the absolute jitter.

EXAMPLE 6   For $f_0 = 1$GHz and $\mathcal{L}_0 = -130$dBc/Hz, the RMS N-period jitter is independent of N and equal to 2.25ps.

$1/f^2$ Phase Noise Profile

In this section, the N-period jitter corresponding to a $1/f^2$ characteristic of the phase PSD is derived. This corresponds to the phase noise of a free-running oscillator with internal white noise sources. Assuming

$$\mathcal{L}(f) = \frac{\mathcal{L}_1 f_1^2}{f^2} \tag{3.53}$$

as shown in Figure 3.11 and substituting the expression of the phase noise into Equation 3.49, the N-period jitter is:

$$\sigma_{p(N)}^2 = \frac{8\mathcal{L}_1 f_1^2}{\omega_0^2} \int_0^{+\infty} \frac{\sin^2(\pi f N/f_0)}{f^2}\,df = \frac{2\mathcal{L}_1 f_1^2 N}{\pi f_0^3} \int_0^{+\infty} \frac{\sin^2(v)}{v^2}\,dv. \tag{3.54}$$

This integral can be calculated per parts, and, using the definite integral $\int_0^{+\infty} \frac{\sin(v)}{v}\,dv = \pi/2$, we find finally the following expressions for the N-period and period jitter:

$$\sigma_{p(N)} = \sqrt{\frac{\mathcal{L}_1 f_1^2}{f_0^3} N} \tag{3.55}$$

$$\sigma_{p} = \sqrt{\frac{\mathcal{L}_1 f_1^2}{f_0^3}}. \tag{3.56}$$

It can be seen that, for a $1/f^2$ phase noise, the RMS N-period jitter increases indefinitely with the square root of the number of cycles N, and the long-term jitter is infinite. An experimental proof of Equation 3.56 has been provided in [29]. Note also that, as stated above, even though the phase noise has infinite power at low frequencies, the computation of the N-period jitter leads to a finite result even if the integration is carried out down to zero frequency offset.

EXAMPLE 7   For an oscillator running at f_0=1GHz, having a $1/f^2$ spectrum with -130dBc/Hz at 1MHz from the carrier, the RMS period jitter is equal to 10fs.

Simplified PLL Phase Noise Profile

We now reconsider a phase noise profile approximating a PLL spectrum such as the one shown in Figure 3.12 and expressed in Equation 3.37, repeated here for convenience:

$$\mathcal{L}(f) = \frac{\mathcal{L}_0}{1 + (f/f_{3dB})^2}. \tag{3.57}$$

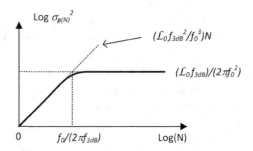

Figure 3.14 N-period jitter profile for a simplified PLL spectrum.

By taking the inverse Fourier transform:

$$R_\varphi(\tau) = \mathcal{L}_0 \pi f_{3dB}\, e^{-2\pi f_{3dB}|\tau|}. \tag{3.58}$$

From here, and using Equation 3.45 the following expression can be derived for the N-period jitter:

$$\sigma_{\mathbf{p}(N)} = \sqrt{\frac{\mathcal{L}_0 f_{3dB}}{2\pi f_0^2}\,(1 - e^{-2\pi f_{3dB}N/f_0})}. \tag{3.59}$$

The shape of $\sigma_{\mathbf{p}(N)}^2$ versus N presents two asymptotes, one for low values of N and one for large values of N, as illustrated in Figure 3.14 in a log–log scale. Note that, in this graph, N essentially represents time.

For large N values, the N-period jitter is independent for N and equal to:

$$\lim_{N\to+\infty} \sigma_{\mathbf{p}(N)} = \sigma_{lt} = \sqrt{\frac{\mathcal{L}_0 f_{3dB}}{2\pi f_0^2}}. \tag{3.60}$$

For small values of N, a simple expression can be found considering that in real PLLs the bandwidth must be smaller – usually at least 10 times smaller – than the input reference frequency. Therefore, also in case of low feedback divider ratios, we can assume f_{3dB} to be much smaller than the output frequency f_0. Starting from Equation 3.59 and using the approximation $e^{-x} = 1 - x$ for small x, the N-period jitter can be written as:

$$\sigma_{\mathbf{p}(N)} = \sqrt{\frac{\mathcal{L}_0 f_{3dB}^2}{f_0^3} N} \tag{3.61}$$

showing that the RMS N-period jitter initially grows with the square root of the number of cycles N. It can be also seen that this expression is identical to Equation 3.55 obtained in the case of a pure $1/f^2$ phase noise profile.

This result indicates that the noise on the edges produced by the locked VCO is equal to that of a free-running VCO until the PLL feedback kicks in and takes control. At this point, the jitter is limited by the loop with an RMS value of σ_{lt}. In the absence of feedback, $\sigma_{\mathbf{p}(N)}$ would grow indefinitely according to the initial slope.

By setting $N = 1$ in Equation 3.61, the period jitter is obtained as:

$$\sigma_p = \sqrt{\frac{\mathcal{L}_0 f_{3dB}^2}{f_0^3}}. \tag{3.62}$$

While the long-term jitter depends on the carrier frequency f_0 and on the total area of the phase noise, the period jitter depends on the carrier frequency and on the position of the $1/f^2$ slope in the PSD, which, in PLLs, is determined mainly by the noise performance of the VCO.

It is interesting to investigate the relation between period and long term jitter. Taking Equations 3.60 and 3.62, the ratio between the RMS values of long term and period jitter is:

$$\frac{\sigma_{lt}}{\sigma_p} = \sqrt{\frac{f_0}{2\pi f_{3dB}}} = \sqrt{\frac{\tau_{loop}}{T_0}} \tag{3.63}$$

where τ_{loop}, defined as $1/(2\pi f_{3dB})$, is the time constant of the PLL loop. It can be seen that the difference between long-term and period jitter grows larger the more slowly the PLL loop reacts to changes in the excess phase at its input. Since the period jitter is mainly determined by the noise performance of the VCO, Equation 3.63 says that, for a given VCO, the long-term jitter increases with the square root of the loop time constant. Reducing the loop time constant by two, meaning doubling the PLL bandwidth, reduces the long-term jitter by a factor of $\sqrt{2}$. Note that this is valid only if the VCO is the main contributor to the output noise of the PLL. If this is not the case, increasing the PLL bandwidth might even degrade the long term jitter, by allowing more noise from the other blocks to reach the output.

Another interesting expression of the long-term jitter can be obtained starting from Equation 3.60:

$$\sigma_{lt} = \sqrt{\frac{\mathcal{L}_0 f_{3dB}^2}{2\pi f_{3dB} f_0^2}} \tag{3.64}$$

where it becomes clear that, in a first approximation, for a given VCO (fixed $\mathcal{L}_0 f_{3dB}^2$), the RMS long-term jitter improves with the square root of the bandwidth of the PLL.

EXAMPLE 8   For a PLL with output frequency $f_0 = 200\text{MHz}$, an in-band phase noise of -120dBc/Hz and 1MHz bandwidth, we have an RMS period jitter of 350fs and an RMS long-term jitter of 2ps. If the bandwidth is reduced to 100kHz, keeping the same VCO, the in-band noise level increases to -100dBc/Hz, in this case the RMS period jitter is 353fs (almost no change) and the RMS long-term jitter increase to 6.3ps ($\sqrt{10}$ increase).

$1/f^3$ Phase Noise Profile

In this section the N-period jitter corresponding to a $1/f^3$ characteristic of the phase noise is derived. This phase noise profile is typical of the low frequency behavior of a free-running oscillator with internal flicker noise sources. Assuming

$$\mathcal{L}(f) = \frac{\mathcal{L}_1 f_1^3}{f^3} \tag{3.65}$$

and substituting the expression into Equation 3.49:

$$\sigma_{\mathbf{p}(N)}^2 = \frac{8\mathcal{L}_1 f_1^3}{\omega_0^2} \int_0^{+\infty} \frac{\sin^2(\pi f N/f_0)}{f^3} \, df. \tag{3.66}$$

For small values of f, the integrand varies as $1/f$, which would lead to a non-converging integral. Therefore we assume that the flicker noise extends towards low frequencies only down to a positive frequency f_{min}. For $f < f_{min}$ the phase noise PSD is assumed to be zero. The integral in Equation 3.66 can be expressed as:

$$\int_{f_{min}}^{+\infty} \frac{\sin^2(\pi f N/f_0)}{f^3} \, df = \frac{(\pi N/f_0)^2}{2} \left[\frac{\sin^2(f_{min}\pi N/f_0)}{(f_{min}\pi N/f_0)^2} \right.$$

$$\left. + \frac{\sin(2f_{min}\pi N/f_0)}{f_{min}\pi N/f_0} - 2\mathrm{Ci}(2f_{min}\pi N/f_0) \right] \tag{3.67}$$

where the function Ci is the cosine integral defined as:

$$\mathrm{Ci}(x) := -\int_x^{+\infty} \frac{\cos(t)}{t} \, dt = \gamma + \ln(x) + \int_0^x \frac{\cos(t) - 1}{t} \, dt \tag{3.68}$$

where ln is the natural logarithm and $\gamma \approx 0.5772$ is the Euler–Mascheroni constant (see [30])[3]. Substituting these expressions in Equation 3.66 and assuming $f_{min}\pi N/f_0 \ll 1$, the following expression for the N-period jitter can be found:

$$\sigma_{\mathbf{p}(N)}^2 = \frac{\mathcal{L}_1 f_1^3 N^2}{f_0^4} [3 - 2\mathrm{Ci}(2f_{min}\pi N/f_0)]. \tag{3.69}$$

For x tending to zero, the function $\mathrm{Ci}(x)$ can be approximated by $\ln(x) + \gamma$, therefore, for values of $N \ll f_0/(2\pi f_{min})$:

$$\sigma_{\mathbf{p}(N)}^2 = \frac{\mathcal{L}_1 f_1^3 N^2}{f_0^4} [3 - 2\gamma - 2\ln(2\pi f_{min} N/f_0)]. \tag{3.70}$$

Since $\ln(x)$ is a very weak function of x, the expression above predicts that the RMS N-period jitter increases almost linearly with the number of cycles N. The same expression has also been derived in [31] with a different approach, and a very similar one can be found in [32], albeit with different notation.

It is important to note that, in practice, the assumption $N \ll f_0/(2\pi f_{min})$ holds very well, so that it does not limit the validity of the formulas derived here. Indeed, to measure the N-period jitter and have significant statistics on it, it is necessary in practice to observe the clock for a time much longer than N periods. It can be argued that the lowest frequency component detectable in this experiment, the f_{min} in the expressions above, is the reciprocal of the observation time. Combining these two considerations, it follows that $1/f_{min} \gg N/f_0$, which is identical to the assumption before, apart from an irrelevant factor 2π.

---

[3] The function Ci is named "cosint" in Matlab.

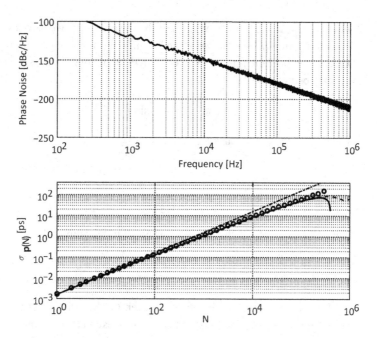

Figure 3.15 $1/f^3$ phase noise profile and corresponding N-period jitter. The circles are values derived from simulation, the dashed line is the RMS N-period jitter obtained using Equation 3.67, the solid line is the simplified Expression 3.70, and the dash-dotted line represents a linear dependency.

The formulas derived in this section have been verified with numerical simulation. Using the methods that will be described in Chapter 10, one million clock cycles of a 10MHz clock have been numerically generated, with a phase noise of -120dBc/Hz at 1kHz and pure $1/f^3$ profile (see Figure 3.15, top). The corresponding N-period jitter has been computed based on the time domain samples of the jitter and compared to the analytical expressions found above, where f_{min} has been chosen as the reciprocal of the total observation time (in this particular case $f_{min} = 10e6/1e6 = 10$Hz). As can be seen in Figure 3.15, the agreement between the simulation and the analytical expression is very good up to a lag of about one tenth of the total number of cycles. The dash-dotted line represents a pure linear relationship between RMS jitter and number of cycles N and shows that the increase of the RMS value is slower than linear, due to the effect of the logarithm in Equation 3.70.

3.3   Spectral Spurious Tones and Jitter

Recalling Section 3.1.4, the presence of symmetrical tones at frequency f_m from the carrier f_0 in the voltage spectrum of a clock signal indicates a deterministic modulation of the excess phase of the form:

$$\varphi(t) = \Phi_m \sin(2\pi f_m t + \phi_0) \tag{3.71}$$

where $\Phi_m = 2 \cdot 10^{(\mathrm{SCR}/20)}$, with SCR the difference between tones and carrier levels in dB (see Section 3.1.4), and ϕ_0 a random initial phase. From here it is quite straightforward to compute the jitter associated with discrete tones in the spectrum. From Equation 3.5 the absolute jitter is:

$$\mathbf{a}_k = -\frac{\Phi_m}{\omega_0} \sin\left(\frac{2\pi f_m k}{f_0} + \phi_0\right). \tag{3.72}$$

The peak values of the absolute jitter are:

$$\rho_{\mathbf{a}}^+ = \rho_{\mathbf{a}}^- = \max_k |\mathbf{a}_k| = \frac{\Phi_m}{\omega_0} \tag{3.73}$$

$$\rho\rho_{\mathbf{a}} = 2\frac{\Phi_m}{\omega_0}. \tag{3.74}$$

The RMS value of the absolute jitter is equal to the amplitude of the sinusoid divided by $\sqrt{2}$. The result is:

$$\sigma_{\mathbf{a}} = \frac{1}{\sqrt{2}} \frac{\Phi_m}{\omega_0}. \tag{3.75}$$

With an expression for the absolute jitter, the N-period jitter can be easily computed using Equation 2.12. After some algebra, the result can be expressed as:

$$\mathbf{p}_k(N) = -2\frac{\Phi_m}{\omega_0} \cos\left(\frac{2\pi f_m k}{f_0} + \phi_0 + \frac{\pi f_m N}{f_0}\right) \sin\left(\frac{\pi f_m N}{f_0}\right). \tag{3.76}$$

The peak values are:

$$\rho_{\mathbf{p}(N)}^+ = \rho_{\mathbf{p}(N)}^- = \max_k |\mathbf{p}_k(N)| = 2\frac{\Phi_m}{\omega_0} \left|\sin\left(\frac{\pi f_m N}{f_0}\right)\right| \tag{3.77}$$

$$\rho\rho_{\mathbf{p}(N)} = 4\frac{\Phi_m}{\omega_0} \left|\sin\left(\frac{\pi f_m N}{f_0}\right)\right| \tag{3.78}$$

and the RMS value:

$$\sigma_{\mathbf{p}(N)} = \sqrt{2}\frac{\Phi_m}{\omega_0} \left|\sin\left(\frac{\pi f_m N}{f_0}\right)\right|. \tag{3.79}$$

Note how the RMS and peak N-period jitter have a periodical behavior in N, with period equal to f_0/f_m.

3.4   Superposition of Different Spectral Components

In this section it will be demonstrated that, in the case of a phase noise PSD made up of different components, the total RMS N-period jitter is the square root of the sum of the squared RMS N-period jitter of each component. In other words, the RMS N-period jitter adds quadratically.

Assume the phase noise profile $\mathcal{L}$ can be expressed as the sum of k different components $\mathcal{L}_1(f), \ldots, \mathcal{L}_k(f)$, then:

$$\mathcal{L}(f) = \sum_{i=1}^{k} \mathcal{L}_i(f). \tag{3.80}$$

Table 3.1 General relationships between variance of jitter and phase noise.

Jitter	Symbol	As a Function of $\mathcal{L}(f)$
Absolute	$\sigma_{\mathbf{a}}^2$	$\dfrac{2}{\omega_0^2} \displaystyle\int_0^{+\infty} \mathcal{L}(f)\, df$
Period	$\sigma_{\mathbf{p}}^2$	$\dfrac{8}{\omega_0^2} \displaystyle\int_0^{+\infty} \mathcal{L}(f)\sin^2(\pi f/f_0)\, df$
N-period	$\sigma_{\mathbf{p}(N)}^2$	$\dfrac{8}{\omega_0^2} \displaystyle\int_0^{+\infty} \mathcal{L}(f)\sin^2(\pi f N/f_0)\, df$
Allan Deviation	$\sigma_{\Delta y}^2\left(\dfrac{N}{f_0}\right)$	$\left(\dfrac{2}{\pi N}\right)^2 \displaystyle\int_0^{+\infty} \mathcal{L}(f)\sin^4(\pi f N/f_0)\, df$

Table 3.2 Variance of jitter for specific phase noise profiles. The variance of the period jitter $\sigma_{\mathbf{p}}^2$ can be obtained from $\sigma_{\mathbf{p}(N)}^2$ with $N=1$.

	$\mathcal{L}(f)$	$\sigma_{\mathbf{a}}^2$	$\sigma_{\mathbf{p}(N)}^2$
Flat	$\mathcal{L}_0$	$\dfrac{\mathcal{L}_0}{4\pi^2 f_0}$	$\dfrac{\mathcal{L}_0}{2\pi^2 f_0}$
$1/f^2$	$\dfrac{\mathcal{L}_1 f_1^2}{f^2}$	$\dfrac{\mathcal{L}_1 f_1^2}{2\pi^2 f_0^2 f_{min}}$	$\dfrac{\mathcal{L}_1 f_1^2}{f_0^3}N$
$1/f^3$	$\dfrac{\mathcal{L}_1 f_1^3}{f^3}$	$\dfrac{\mathcal{L}_1 f_1^3}{(2\pi f_0 f_{min})^2}$	$\dfrac{\mathcal{L}_1 f_1^3 N^2}{f_0^4}[3 - 2\gamma - 2\ln(2\pi f_{min} N/f_0)]$
Simple PLL	$\dfrac{\mathcal{L}_0}{1 + (f/f_{3dB})^2}$	$\dfrac{\mathcal{L}_0 f_{3dB}}{4\pi f_0^2}$	$\dfrac{\mathcal{L}_0 f_{3dB}}{2\pi f_0^2}(1 - \exp(-2\pi f_{3dB}N/f_0))$

Table 3.3 Relationship between jitter and SCR.

Jitter	Symbol	As a Function of SCR
RMS Absolute	$\sigma_{\mathbf{a}}$	$\dfrac{\sqrt{2}}{\omega_0} 10^{(SCR_{dBc}/20)}$
RMS Period	$\sigma_{\mathbf{p}}$	$2\sigma_{\mathbf{a}}\lvert\sin(\pi f_m/f_0)\rvert$
RMS N-Period	$\sigma_{\mathbf{p}(N)}$	$2\sigma_{\mathbf{a}}\lvert\sin(\pi f_m N/f_0)\rvert$
peak Absolute	$\rho_{\mathbf{a}}$	$\sqrt{2}\sigma_{\mathbf{a}}$
peak Period	$\rho_{\mathbf{p}}$	$\sqrt{2}\sigma_{\mathbf{p}}$
peak N-Period	$\rho_{\mathbf{p}(N)}$	$\sqrt{2}\sigma_{\mathbf{p}(N)}$

Due to the linearity of the Fourier transform:

$$R_\varphi(\tau) = \sum_{i=1}^{k} R_{\varphi,i}(\tau),$$ (3.81)

therefore we can write:

$$R_\varphi(0) - R_\varphi(\tau) = \sum_{i=1}^{k} [R_{\varphi,i}(0) - R_{\varphi,i}(\tau)],$$ (3.82)

and, finally, using Equation 3.45:

$$\sigma^2_{\mathbf{p}(N)} = \sum_{i=1}^{k} \sigma^2_{\mathbf{p}(N),i}.$$ (3.83)

3.5   Summary of Mathematical Relationships Between Jitter and Phase Noise

In this section, the most important formulas that relate jitter to phase noise or Spurious to Carrier Ratio (SCR) (see Section 3.1.4) are summarized in tabular form. Table 3.1 reports the general formulas used to calculate the variance of absolute, period and N-period jitter, as well as the Allan Deviation (see Section 9.3.4), starting from the phase noise $\mathcal{L}(f)$. Table 3.2 shows the results for the special cases of phase noise having a flat, a $1/f^2$, a $1/f^3$ and a low-pass profile. Finally Table 3.3 lists the formulas used to calculate RMS and peak values of absolute, period and N-period jitter in the case of discrete tones in the spectrum, starting from the SCR.

4 Jitter and Phase Noise in Circuits

This chapter will offer an introduction to how jitter and phase noise are generated in the most common oscillator topologies. Starting with the treatment of basic circuits like a current charging a capacitor, or an inverter, we will continue by analyzing jitter and phase noise generation in ring, relaxation, and LC oscillators. Both linear time invariant and linear time variant approaches will be illustrated.

The chapter will finish with an overview of how jitter is "transformed" by two of the most common frequency processing circuits: a digital frequency divider and an ideal frequency multiplier.

4.1 Jitter in Basic Circuits

The next sections provide an explanation, both qualitative and quantitative, of how jitter manifests itself in basic oscillator circuits.

4.1.1 Noisy Current Charging a Capacitor

As mentioned already in Section 1.3, in integrated circuit design a central element is a current loading a capacitor. Digital gates, clock buffers, delay elements of relaxation and ring oscillators are just few examples of where this element can be found. Due to its ubiquity and importance, it is worth analyzing in some detail from the standpoint of jitter generation. In the following we will revisit the example provided in Section 1.3, and treat it with a complete mathematical analysis. The case of a PMOS transistor charging a capacitor is shown in Figure 4.1. The case of an NMOS discharging the capacitor is identical.

The goal of this analysis is to derive an expression for the variation of the time instant at which the charging voltage crosses a given threshold V_0 (see Figure 4.1). The first step is to recognize that the resulting time deviation is due to two different components. The first is the effect of the noise current i_n injected into the capacitor by the PMOS during the charging transient. The second is due to the fact that, even if the capacitor is considered discharged at time zero, there is voltage noise present on it due to the circuitry which is used to keep it discharged. This noise can be considered as a random initial vertical offset which will affect the threshold crossing time.

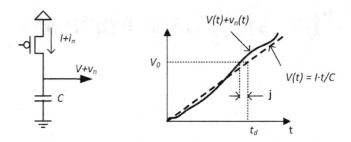

Figure 4.1 PMOS current source charging a capacitor.

Starting with the first component, the voltage noise due to the noisy current is given by:

$$v_{n1}(t) = \frac{1}{C} \int_0^t i_n(\tau) d\tau. \tag{4.1}$$

It is well known from the theory of stochastic processes (see [1]) that the integral over time of stationary white noise is the so-called *Wiener process*, which belongs to the category of random walks and is not stationary. Indeed the distance traveled by the voltage $v_{n1}(t)$ from the starting point is zero on average, but its variance increases linearly with time t. For t going to infinity, the voltage $v_{n1}(t)$ will show an infinite variance. Fortunately, in the case under discussion here, the process of integration of the noisy current is limited to a finite observation time, which is the time t_d taken by the voltage ramp to reach the threshold V_0. What happens after this time is not of interest. By considering the integration only over a finite observation time $[0, t_d]$, the result is again a stationary process and the classic theory of stationary processes in linear systems can be used.

To derive an expression for the variance of $v_{n1}(t)$ at time t_d, it can be observed that the integral in Equation 4.1 can be written as a convolution (see [33]):

$$v_{n1}(t_d) = \frac{1}{C} \int_{-\infty}^{+\infty} i_n(\tau) w(t_d - \tau) d\tau \tag{4.2}$$

where $w(\cdot)$ is a window function, assuming value 1 in the interval $[0, t_d]$ and zero outside, and v_{n1} indicates the voltage noise at the end of the observation time $[0, t_d]$. This is the integral form of a linear filter with impulse response $w(t)$ applied to the noisy current. Therefore the relation between the PSD of the current noise and of the voltage noise is:

$$S_{v_{n1}}(f) = \frac{S_{i_n}(f)}{C^2} |W(f)|^2 = \frac{S_{i_n}(f)}{C^2} \frac{\sin^2(\pi f t_d)}{(\pi f)^2} \tag{4.3}$$

where $W(f)$ is the Fourier transform of $w(t)$. From the PSD, the variance can be computed by integration over the whole frequency axis. Assuming that the current i_n is a white noise with $S_{i_n}(f) = S_{i_n}$ independent of f, the variance of the noise voltage is:[1]

$$\sigma_{v_{n1}}^2 = \int_{-\infty}^{+\infty} S_{v_{n1}}(f) df = \frac{S_{i_n}}{C^2} t_d. \tag{4.4}$$

---

[1] Note that in [33] the result has a 2 at the denominator. The reason is that here the PSD is considered two-sided, while [33] considers it one-sided.

Starting from this expression, if the PMOS is in saturation, the PSD of the current noise is $S_{i_n} = 2kT\gamma g_m$ and the variance of the noise voltage is:

$$\sigma_{v_{n1}}^2 = \frac{2kT\gamma g_m t_d}{C^2}. \tag{4.5}$$

To a first approximation, the timing error (jitter) is obtained by dividing the amplitude error by the slope of the signal. In this case, the slope is I/C so that the variance of the jitter can be written as:

$$\sigma_{j1}^2 = \frac{2kT\gamma g_m t_d}{I^2}. \tag{4.6}$$

Note how the variance of jitter grows linearly with the observation time, for a given current I.

Since $V_0 = I t_d/C$, the jitter at the instant t_d can be finally expressed as:

$$\sigma_{j1}^2 = \frac{4kT\gamma t_d^2}{V_0 C(V_{GS} - V_{th})} \tag{4.7}$$

where for g_m the expression $2I/(V_{GS} - V_{th})$ has been used assuming the PMOS is in saturation, with V_{GS} and V_{th} the gate-source and threshold voltage respectively. Note that if the device is in weak-inversion $g_m = I/(nkT/q)$ (n is the subthreshold slope factor, typically between 1 and 2) which is higher than in saturation, for the same current I.

The second jitter component is due to the initial voltage noise present on the capacitor. It is well known that, independent of the output impedance of the circuit used to keep the capacitor in a discharged state, the voltage noise has a variance given by:

$$\sigma_{v_{n2}}^2 = \frac{kT}{C}. \tag{4.8}$$

This initial voltage noise represents an amplitude offset which is carried unchanged along the whole charging transient. To get to the corresponding jitter, it must be divided by the by the slope of the signal, so that:

$$\sigma_{j2}^2 = \frac{kT\, t_d^2}{V_0^2 C}. \tag{4.9}$$

Since the mechanisms generating the two jitter contributions are uncorrelated, the variance of the total jitter is the sum of the variances of the two components. This leads to the following final expression for the RMS of the jitter relative to the delay t_d:

$$\left(\frac{\sigma_j}{t_d}\right)^2 = \frac{kT}{V_0 C} \left(\frac{4\gamma}{V_{GS} - V_{th}} + \frac{1}{V_0}\right). \tag{4.10}$$

EXAMPLE 9   Consider the case of an inverter in $1\,\text{V}$ CMOS technology charging a $10\,\text{fF}$ capacitor. If we set V_0 at half supply ($0.5\,\text{V}$), consider a moderate overdrive $V_{GS} - V_{th} = 200\,\text{mV}$, and take for γ the common value of $2/3$, at 300 degree Kelvin, the relative jitter is $\sigma_j/t_d = 0.11\%$ of the delay. If $t_d=100\,\text{ps}$, the RMS value of the jitter introduced by this inverter is about $0.11\,\text{ps}$. Note also that the contribution due to the initial noise voltage level is much smaller than the contribution of the device.

It is worth summarizing the assumptions that have been made to get to this simple expression. First, the charging current has been considered to be constant over the whole time (not generally true due to the finite output resistance of the PMOS). Second, the noise of the circuit driving the gate of the PMOS and the flicker noise of the PMOS itself have been neglected. Third, the formulas for the g_m and the noise current are based on a first-order MOS model in saturation and strong inversion.

Despite all these assumptions, Equation 4.10 can be used for a first-order assessment, suggesting a few ways to improve jitter performance of this kind of current driving stages for a given delay t_d. First, the overdrive voltage of the current source $V_{GS} - V_{th}$ should be maximized. Second, the threshold V_0 should be as large as possible, which, for a given delay, requires increasing the nominal current. Third, the load capacitor C should be made as big as possible, which again leads to an increase in the charging current. Apart from the last technique, the first two are constrained by the technology used. The overdrive and the threshold can not be made larger than the supply of the circuit (V_{DD}), so that the following lower bound on achievable relative jitter can be found (4γ has been approximated with the number 3):

$$\frac{\sigma_j}{t_d} \geq \frac{2}{V_{DD}} \sqrt{\frac{kT}{C}}. \tag{4.11}$$

4.1.2   Jitter of a CMOS Inverter

A very fundamental element in modern integrated systems is the CMOS inverter, shown in Figure 4.2. This element is used not only in the context of digital processing, but also as part of clock distribution trees in digital systems or as a delay element in a delay line or ring oscillator. It is thus of great relevance to derive an expression for the jitter introduced by it. A qualitative introduction was given in Section 1.4. Here we will analyze it from a more quantitative perspective. Assuming that the inverter input voltage has infinite slope, each charge or discharge of the load capacitor is essentially performed by the either PMOS or NMOS device only, respectively. The other device is instantaneously turned off and does not contribute any current. Therefore, the simple model derived in the previous section can be applied as a first approximation. By using Equation 4.10 with $V_{GS} = V_{DD}$ and assuming that the switching threshold is at half

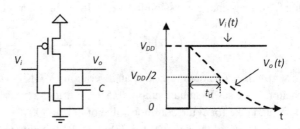

Figure 4.2 CMOS inverter and input–output waveforms in the approximation leading to Equation 4.12.

supply, $V_0 = V_{DD}/2$, the variance of the jitter relative to the inverter delay can be written as (see also [33]):

$$\left(\frac{\sigma_j}{t_d}\right)^2 = \frac{4kT}{CV_{DD}^2}\left(1 + \frac{2\gamma V_{DD}}{V_{DD} - V_{th}}\right). \tag{4.12}$$

It can be seen that, for a given supply and a given capacitive load, the RMS value of the jitter is proportional to the inverter delay. The stronger the inverter, the smaller the jitter. Note that Equation 4.12 does not take into account the effect of the finite input slope, which might further degrade the jitter. It also assumes that the MOS device is in saturation at least till the switching threshold is reached. This assumption is a simplification of the real behavior, where the MOS device, depending on technology and supply voltage values, might enter the linear region and charge or discharge the output load with a non-constant current.

EXAMPLE 10   Consider the case of a CMOS inverter in 1V technology driving a 10fF load. If we take $\gamma = 2/3$ and a MOS threshold voltage of 0.4V, at 300 degrees Kelvin the relative jitter is $\sigma_j/t_d = 0.23\%$ of the delay. If the inverter is dimensioned to give a $t_d=100$ps, the RMS value of the jitter introduced is about 0.23ps.

In addition to the thermal noise of the MOS devices, supply noise is also a major cause of jitter in a CMOS inverter, since this single-ended structure has a very low power supply rejection. In the context of a complex chip with a lot of switching activity on the supplies, this source of noise can be dominant and the corresponding jitter be much larger than the thermal one.

Considering the expression for the delay $t_d = V_0 C/I$, the effect of a varying supply is twofold. First, the inverter threshold voltage V_0 changes, since it is normally proportional to the supply. Second, the change in supply also causes a change in the charging current I. A simple first-order expression for the jitter due to supply noise can be easily derived. Assuming that V_0 is proportional to V_{DD}, the change in the delay, Δt_d, can be decomposed in the contribution of the change of V_0 and of I in the following way:

$$\frac{\Delta t_d}{t_d} = \frac{\Delta V_0}{V_0} - \frac{\Delta I}{I} = \frac{\Delta V_{DD}}{V_{DD}} - \frac{\Delta I}{I}. \tag{4.13}$$

Noting that V_{DD} is the gate-source voltage of the MOS device charging (or discharging) the load capacitor, the change in current due to the change in V_{DD} is given by the transconductance g_m of the MOS device, $\Delta I = g_m \Delta V_{DD}$. To obtain a simple expression, we will assume that the device is in saturation at least till the output voltage reaches the switching threshold, an assumption which might not always hold. Using Equation 4.13 with $V_{GS} = V_{DD}$ and $g_m = 2I/(V_{DD} - V_{th})$ then, the delay variation can be written as:

$$\frac{\Delta t_d}{t_d} = -\left(\frac{V_{DD} + V_{th}}{V_{DD} - V_{th}}\right) \cdot \frac{\Delta V_{DD}}{V_{DD}}. \tag{4.14}$$

EXAMPLE 11   Consider the case of the previous example, a change of 1% in supply voltage results in a change of $(1 + 0.4)/(1 - 0.4) \cdot 1\% = 2.3\%$ of the delay, which is much larger than the RMS jitter induced by the self-generated thermal noise.

Two observations can be made about the factor relating the relative change in V_{DD} to the relative change of the delay in Equation 4.14. First, this factor is negative, meaning that, in case of a positive ΔV_{DD}, the increased delay due to an increase in the threshold level is more than compensated for by the increase in charging current. Secondly, the absolute value of the factor approaches 1 only for $V_{DD} \gg V_{th}$. For V_{DD} approaching V_{th} the factor increases asymptotically, indicating a severe degeneration of the jitter performance of the inverter. In this case, however, the MOS devices enter weak inversion and another transistor model should be used (see, e.g., [34]).

In modern technologies, very short channel length devices enter *velocity saturation* for large overdrive voltage $V_{GS} - V_{th}$, as in the case of a minimum length inverter with fast switching signal at its input. The dependence of the drain current on the overdrive voltage tends to be more linear, rather than quadratic, so that g_m can be expressed as $\alpha I / (V_{GS} - V_{th})$, with $1 < \alpha < 2$. Using this expression, Equation 4.14 is modified as follows:

$$\frac{\Delta t_d}{t_d} = -\left[\frac{(\alpha - 1)V_{DD} + V_{th}}{V_{DD} - V_{th}}\right] \cdot \frac{\Delta V_{DD}}{V_{DD}}. \tag{4.15}$$

For $\alpha = 2$ (quadratic dependence of drain current on overdrive voltage) this equation reduces to Equation 4.14. If V_{DD} is affected by deterministic noise, such as modulation, IR drop, or switching noise, the resulting jitter is DJ and its peak value ρ_j is related to the peak value $\rho_{V_{DD}}$ of V_{DD} by:

$$\frac{\rho_j}{t_d} = -\left[\frac{(\alpha - 1)V_{DD} + V_{th}}{V_{DD} - V_{th}}\right] \cdot \frac{\rho_{V_{DD}}}{V_{DD}}. \tag{4.16}$$

If V_{DD} is affected by random thermal noise with PSD $S_{V_{DD}}(f)$, the jitter is RJ with PSD:

$$S_j(f) = \left[\frac{(\alpha - 1)V_{DD} + V_{th}}{V_{DD} - V_{th}}\right]^2 \left(\frac{t_d}{V_{DD}}\right)^2 \cdot S_{V_{DD}}(f) \tag{4.17}$$

and an equation similar to Equation 4.16 will relate the RMS values of the V_{DD} noise and of the jitter:

$$\frac{\sigma_j}{t_d} = \left[\frac{(\alpha - 1)V_{DD} + V_{th}}{V_{DD} - V_{th}}\right] \cdot \frac{\sigma_{V_{DD}}}{V_{DD}}. \tag{4.18}$$

It must be understood that the previous equations are first-order approximations and are subject to the same assumptions as those made in Section 4.1.1 to compute the charging of the capacitor. Additionally, Equation 4.13 and all those derived from it are obtained by linearization around the nominal value, and therefore are valid only if the relative change of V_{DD} and current I are small. Finally, the effect of the finite output resistance and of the parasitic capacitances of the MOS devices are neglected. For a deeper discussion, see specific papers such as, e.g., [35] and [36].

4.1.3   Jitter of a CMOS Differential Stage

Another basic building block is the CMOS differential stage with resistive load shown in Figure 4.3. Since this block is also widely used in differential clock distribution

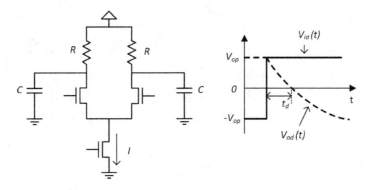

Figure 4.3 Resistive loaded differential pair and input–output waveforms in the approximation leading to expression 4.19. $V_{id}(t)$ and $V_{od}(t)$ are the input and output differential voltages.

networks, ring oscillators or delay lines, it is of relevance to compute an expression of the jitter due to the thermal noise of the devices. In practical implementations, the load resistor R is often implemented using a combination of MOS devices biased in the linear region, but this is not essential for jitter analysis. A thorough analysis of the variation of the delay of this stage due to thermal noise has been carried out in [37] and [33]. The time delay t_d is in this case defined as the time difference between the zero crossing points of the differential input and output voltages. The general approach is the same as for the single-ended CMOS inverter. The differential amplitude noise at the zero crossing point is computed first, and then divided by the slope of the output signal to derive the timing error. The three sources of noise are the load resistors, the devices of the differential pair and the tail device. The thermal noise of the resistors and of the differential pair devices of the left and right half circuits are uncorrelated, therefore clearly contributing to the noise on the differential output. The tail current noise is injected into a common mode node and it could be erroneously argued that its contribution to the noise at the differential output must be zero. In reality its contribution is not zero. Indeed, with reference to Figure 4.4, assuming the tail current to be initially completely switched to the left branch (Phase 1), the left output node is pulled down and its noise voltage is the contribution of the noise from the load resistor and from the tail current. Assuming, for simplicity, an instantaneous switching of the current into the right branch (Phase 2), the left output starts moving up, carrying the memory of the tail current noise previously accumulated. The right branch starts to be pulled down by the tail current, and the noise of the tail is integrated into the load capacitor. Therefore the noise on the left and the noise on the right branches at the time they cross is due to the contribution of the tail noise current over two distinct time intervals, and therefore are not correlated, yielding a net nonzero noise voltage.

Omitting the math, it can be shown (for the details refer to [33]) that the time delay is given by $t_d = RC \ln 2$ and the variance of the jitter of the delay is given by:

$$\left(\frac{\sigma_j}{t_d}\right)^2 = \frac{2kT}{(\ln 2)^2} \frac{1}{CV_{op}^2} \left[1 + \gamma V_{op} \left(\frac{1}{(V_{GS} - V_{th})_t} + \frac{3/4}{(V_{GS} - V_{th})_d}\right)\right] \qquad (4.19)$$

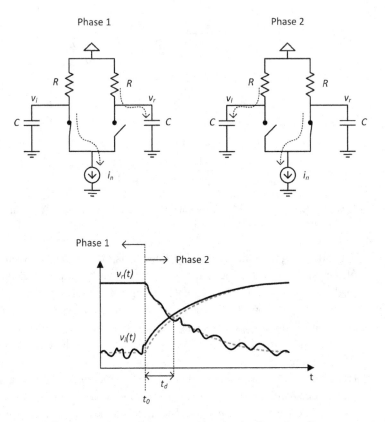

Figure 4.4 Schematic representation of the conversion of noise from the tail current source into jitter. In the bottom picture, the gray dashed line shows the ideal noiseless behavior, while the black solid shows the real noisy behavior.

where I is the tail current, V_{op} is the peak-peak single ended output swing ($V_{op} = IR$), V_{th} denotes the threshold voltage, and $(V_{GS} - V_{th})_t$ and $(V_{GS} - V_{th})_d$ are the effective voltages of the tail and differential pair devices in balanced conditions.

It is interesting to note the similarities between Equations 4.19 and 4.12. For a given delay, both are inversely proportional to the loading capacitor and to the square of the single ended output swing (V_{DD} in the case of the CMOS inverter), and proportional to the ratio of the output swing to the effective voltage of a MOS device. Since the output swing of the differential pair is smaller than V_{DD} and $V_{GS} < V_{DD}$, it is easy to prove (left as an exercise for the reader) that for the same delay and the same load a CMOS inverter has superior jitter performance than the differential stage.

The advantage of the use of a differential stage lies in its high differential mode power supply rejection. If V_{DD} is modulated, both single-ended outputs move up or down by the same amount, in a first approximation, cancelling the effect of the supply noise on the zero crossing time of the differential output. The cost for this is a higher power consumption at lower frequencies (apart from the higher thermal noise jitter), since, in a differential stage, the tail current is always flowing, while in a CMOS inverter, current is drained only during the switching phase.

4.2   Jitter in Oscillators

4.2.1   Ring Oscillators

The delay stages discussed in the previous sections can be used in an inverting ring configuration to obtain a ring oscillator. Ring oscillators are very common building blocks,
widely used as voltage controlled oscillators in phase-locked loops. A qualitative explanation of the mechanism of jitter accumulation in ring oscillators was given in Section
1.4. This section provides the reader with a more quantitative analysis, following the
approach in [37] and [33].

Starting from the jitter produced by each delay stage, it is easy to derive an expression
for the output N-period jitter of a ring oscillator. The first observation is that, in a ring
composed of M delay stages, each with delay t_d, a clock edge needs to go twice through
the whole ring in order to come back to the same location with the same direction (rising
or falling) it had at the beginning. Therefore, the nominal period of oscillation is given
by the sum of $2M$ consecutive delays $t_d(1), t_d(2), \ldots, t_d(2M)$:

$$T_0 = \sum_{k=1}^{2M} t_d(k) = 2Mt_d. \tag{4.20}$$

The variation of the period around its nominal value is given by the contribution of the
variations (jitter) of the $2M$ consecutive delays. The variations in each delay stage are
uncorrelated with each other so that the variance of the sum is given by the sum of the
variances. As a result, the variance of T_0, that is, the variance of the period jitter, σ_p^2 is
given by:

$$\sigma_p^2 = 2M\sigma_j^2 \tag{4.21}$$

where σ_j^2 is the variance of the propagation delay of each stage. Using Equations 4.12
and 4.19 it is now possible to obtain an expression for the period jitter of the ring
oscillator. For an inverter-based ring oscillator, the result is:

$$\sigma_p^2 = \frac{2kT}{f_0 I_{tot} V_{DD}} \left(1 + \frac{2\gamma V_{DD}}{V_{DD} - V_{th}}\right) \tag{4.22}$$

while for a differential stage-based ring oscillator it is:

$$\sigma_p^2 = \frac{2}{\ln 2} kT \frac{M}{f_0 I_{tot} V_{op}} \left[1 + \gamma V_{op} \left(\frac{1}{(V_{GS} - V_{th})_t} + \frac{3/4}{(V_{GS} - V_{th})_d}\right)\right] \tag{4.23}$$

where $f_0 = 1/T_0$ is the nominal oscillation frequency, and I_{tot} is the total current consumption of the complete ring oscillator. Note that for a CMOS inverter-based ring
oscillator, I_{tot} is equal to the current of the single stage I, since only one inverter is
switching at a time. For a differential stage ring oscillator, though, each delay stage
draws a constant current I independent whether it is switching or not, so that $I_{tot} = MI$.
Equation 4.22 is in agreement with the expression found in [38], a paper analyzing
the minimum achievable phase noise in oscillators. From Equations 4.22 and 4.23 it is
evident that the inverter-based ring has superior jitter performance compared with the
differential-based ring, as long as only the thermal noise contribution is considered. A

more detailed comparison is presented in Section 4.5. As mentioned before, though, the inverter-based ring has a very high sensitivity to noise on the supply and, in most cases, requires the use of a low noise voltage regulator. It can be also seen how the period jitter of the inverter-based ring is independent of the number of stages M, for a given frequency f_0 and total power consumption I_{tot}, while that of a differential stage ring increases with M. Therefore, a simple guideline to minimize jitter in an inverter-based ring is to increase V_{DD} as much as possible and draw as much current as the budget allows. The number of stages determines the oscillation frequency. To lower jitter in a differential stage ring oscillator, the output swing and the current consumption should be maximized, and the minimum possible number of stages M should be used.

As already shown in Equation 2.13, the N-period jitter is the sum of the jitter over N consecutive periods. If we assume that the dominant noise sources in ring oscillators are white, the jitter values over any of the N periods are uncorrelated and identically distributed, so that the variances add up to give the variance of the N-period jitter. As a result, the RMS value grows with the square root of the number of periods N:

$$\sigma_{\mathbf{p}(N)} = \sqrt{N}\sigma_{\mathbf{p}}. \tag{4.24}$$

When flicker noise is present, however, the slope of jitter versus N changes. Flicker noise is highly correlated and most of its energy is at low frequencies. For this reason, the assumption that the jitter in consecutive periods is uncorrelated does not hold. If a delay stage has a significant flicker noise component, a part of the variation in its delay during one period of oscillation – the one due to flicker noise – is likely to be very similar to the variation in the previous one. For this case, the accumulation of jitter in consecutive periods is approximately a simple linear process, so that the RMS values, rather than the variances, add together. The result is that the RMS value of the N-period jitter grows linearly with N:

$$\sigma_{\mathbf{p}(N)} = N\sigma_{\mathbf{p}}. \tag{4.25}$$

Therefore, the typical plot of $\sigma_{\mathbf{p}(N)}$ versus N for an oscillator (see Figure 4.5) shows an initial increase with $\sqrt{N}$, which turns into a slope of N as soon as the observation interval, and, therefore N, is large enough for the flicker noise energy to be dominant with respect to the white thermal noise. This topic is discussed also in Section 3.2.2 in the context of phase noise.

4.2.2   Relaxation Oscillators

For low-frequency, low-power clock generation applications, a relaxation oscillator is often the preferred choice. Although the particular topology of the circuit might change from implementation to implementation, all relaxation oscillators are based on the same principle. With reference to Figure 4.6, in the *relaxation* phase, a capacitor, which can be floating or grounded, is charged by a constant current. The capacitor voltage is continuously sensed by a comparator. Once the voltage hits a reference voltage (V_H in Figure 4.6), the comparator triggers a regenerative circuit, normally a simple set–reset flip-flop,

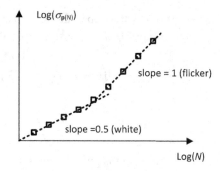

Figure 4.5 Plot of N-period jitter versus N for a free-running oscillator affected by white and flicker noise.

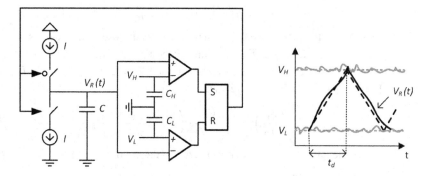

Figure 4.6 Principle block diagram of a relaxation oscillator and relevant waveforms (dashed: ideal; solid: noisy).

which inverts the sign of the charging current (the *regenerative* phase). The capacitor is then discharged linearly until the voltage hits a lower reference voltage (V_L). At this point, the regenerative circuit is triggered again and reverses the current direction, starting the cycle from the beginning. Figure 4.6 shows a principle block diagram of a relaxation oscillator together with the relevant waveforms. In this simplified model, the period jitter can be computed as follows. The uncertainty over the period is given by the sum of the uncertainties over the duration of the rising and falling ramps. Each of these is, in turn, given by two components. The first one is due to the noise on the charging current, as already explained in Section 4.1.1, and the second is the noise present on the reference voltages. The contribution by the noisy current is equal to the amplitude noise on the ramp at time t_d divided by the slope of the ramp itself I/C. Using Equation 4.4, therefore, the jitter contribution due to the ramp noise is:

$$\sigma^2_{j,\text{ramp}} = \frac{S_i}{C^2} t_d \Big/ \left(\frac{I}{C}\right)^2 = \frac{S_i}{I^2} t_d \tag{4.26}$$

where S_i is the PSD of the current source, assumed to be white. This contribution has to be counted twice, since there are two ramps and we are assuming that the noise is uncorrelated. The contribution due to noise on the references can be calculated similarly, as reference noise divided by the slope. Assuming that the capacitances connected to the high and low references are C_H and C_L respectively, the jitter is given by:

$$\sigma^2_{\mathbf{j},\text{thrH}} = \frac{kT}{C_H} \bigg/ \left(\frac{I}{C}\right)^2 \tag{4.27}$$

and

$$\sigma^2_{\mathbf{j},\text{thrL}} = \frac{kT}{C_L} \bigg/ \left(\frac{I}{C}\right)^2. \tag{4.28}$$

The variance of the period jitter is given by the sum of the variances so calculated, with the ramp variance counted twice. After some manipulation, finally the variance of the period jitter can be written as:

$$\sigma^2_{\mathbf{p}} = \frac{1}{(2f_0\Delta V)^2}\left[kT\left(\frac{1}{C_H} + \frac{1}{C_L}\right) + \frac{S_i}{f_0 C^2}\right] \tag{4.29}$$

with ΔV as the difference between the two thresholds (the peak-peak value of the ramp) and $f_0 = I/(2C\Delta V)$ as the oscillation frequency. If the charging current is generated with MOS current mirrors in saturation, then $S_i = 2kT\gamma g_m = 4kT\gamma I/(V_{GS} - V_{th})$ and Equation 4.29 can be rewritten as:

$$\left(\frac{\sigma_{\mathbf{p}}}{T_0}\right)^2 = \frac{kT}{(2\Delta V)^2}\left(\frac{1}{C_H} + \frac{1}{C_L} + \frac{8\gamma}{C}\frac{\Delta V}{V_{GS} - V_{th}}\right) \tag{4.30}$$

where the variance has been normalized to the oscillation period $T_0 = 1/f_0$.

EXAMPLE 12   Consider the case of a 100MHz relaxation oscillator with a ramp swing of 0.5V, 100fF capacitance C, a current mirror with $V_{GS} - V_{th} = 0.3$V and where the two thresholds of the comparator are blocked with 50fF each. Applying Equation 4.30, the RMS value of the period jitter equals 7.3ps.

The derivation above did not consider the contributions to the period jitter coming from the comparators and from the set–reset flip-flop. While they certainly introduce some additional jitter, the regenerative phase, with fast switching and steep edges, is extremely short compared to the ramp time. For this reason its contribution to the total period jitter should be negligible. For more detailed analyses, which consider particular circuit topologies and also other noise sources, see [39], [40], [41], and [42].

Regarding the N-period jitter in relaxation oscillators, the same considerations as for the ring oscillators hold true and can be repeated.

4.3   Phase Noise in Oscillators

4.3.1   Leeson's Model

One of the first publications addressing the modeling of phase noise in oscillators appeared in 1966 [43]. The model proposed is generally known as Leeson's model, after the author, and, despite its limitations, it is still a reference for all recent works on phase noise modeling.

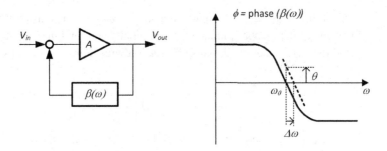

Figure 4.7 Simplified linear system with resonant feedback network $\beta(\omega)$ (left) and corresponding phase response (right).

Leeson's paper considers a linear feedback system as shown in Figure 4.7, where an ideal gain block A is closed in feedback through a resonant network $\beta(\omega)$. The condition for self-sustained oscillation is given by the Barkhausen criterion, stating that the open loop transfer function $A\beta(\omega)$ must be equal to 1. Since A is assumed to be real, the open loop phase $\phi(\omega) = \angle A\beta(\omega)$ is determined solely by β, and the system oscillates at the frequency ω_0 for which $\phi(\omega_0) = 0$.

Figure 4.7 shows the typical phase plot of the resonant network $\beta(\omega)$. The quality factor Q of an oscillator is usually defined as the ratio of the energy stored in the reactive elements to the energy loss per oscillation cycle, multiplied by 2π. In terms of the open loop phase ϕ this definition is equivalent to:

$$Q = \frac{\omega_0}{2}\left|\frac{\partial\phi}{\partial\omega}\right|. \tag{4.31}$$

Following Leeson's approach, noise generated by passive (resistors) or active circuit elements (transistors) disturbs the loop by introducing an additional phase shift $\theta(t)$. Let us assume that $\theta(t)$ is a sinusoidal perturbation at frequency $\omega_0 + \omega_m$. For small ω_m, that is, for perturbation frequencies close to resonance, the loop can sustain oscillation only if the instantaneous frequency of oscillation changes by an amount $\Delta\omega$ so that the condition $\phi(\omega_0 + \Delta\omega) = 0$ is still satisfied. Figure 4.7 shows that the required change in oscillation frequency $\Delta\omega$ is inversely proportional to the slope of the phase curve, thus, recalling Equation 4.31, inversely proportional to the quality factor:

$$\Delta\omega(t) = \frac{\omega_0}{2Q}\theta(t). \tag{4.32}$$

The phase change $\varphi(t)$ at the output of the oscillator is the integral of the frequency change $\Delta\omega$, so that $\theta(t)$ contributes to the output phase multiplied by the factor $\omega_0/(2Q\omega_m)$, and the PSD of $\varphi(t)$ can be written as:

$$S_\varphi(\omega_m) = S_\theta(\omega_m)\left(\frac{\omega_0}{2Q\omega_m}\right)^2. \tag{4.33}$$

The larger the quality factor and the offset frequency ω_m, the smaller the change in the output phase of the oscillator due to internal noise sources.

If the frequency ω_m is large enough so that $\omega_0 + \omega_m$ is outside the resonant region, the value of the feedback branch β is essentially zero and the output phase change φ is equal to θ itself. Combining this observation with Equation 4.33, a suitable expression for the total PSD is:

$$S_\varphi(\omega_m) = S_\theta(\omega_m)\left[1 + \left(\frac{\omega_0}{2Q\omega_m}\right)^2\right]. \tag{4.34}$$

In his paper, Leeson further assumes that the spectrum of the phase perturbation $\theta(t)$ is composed of a white noise floor and a $1/f$ profile at low offset frequencies according to:

$$S_\theta(\omega_m) = \frac{2kTF}{P}\left(1 + \frac{\omega_f}{\omega_m}\right) \tag{4.35}$$

where P is the power consumption of the oscillator, ω_f is the frequency below which the flicker noise component becomes dominant, and F is a noise factor grouping all noise contributors in the loop. Note that if the major part of the power is dissipated in the tank, P can be written as:

$$P = \frac{V_{tank}^2}{2R} \tag{4.36}$$

where V_{tank} is the tank oscillation amplitude and R is the parallel resistance modeling all tank losses.

Since $S_\varphi(\omega_m)$ is the output phase noise of the oscillator $\mathcal{L}(\omega_m)$, combining Equations 4.34 and 4.35, the final model for the phase noise is:

$$\mathcal{L}(\omega_m) = \frac{2kTF}{P}\left(1 + \frac{\omega_f}{\omega_m}\right)\left[1 + \left(\frac{\omega_0}{2Q\omega_m}\right)^2\right] \tag{4.37}$$

as can be seen in Figure 4.8. It is worth pointing out that the Q factor in Leeson's model is the "effective" quality factor of the complete oscillator, not of the tank only. Indeed the Q of a resonator tank disconnected form any other circuitry, also called the *unloaded* Q, is degraded by the losses and parasitic capacitances of the circuitry needed to start and sustain the oscillation. The resulting Q, also called the *loaded* Q, can be much lower than the unloaded one.

EXAMPLE 13   Consider the case of a 1GHz oscillator with loaded $Q = 5$, consuming 1mW in the tank, at room temperature. Assuming realistically $F = 2$, Leeson's model

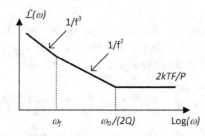

Figure 4.8 Phase noise profile of an oscillator according to Leeson's model.

predicts a phase noise of about -128dBc/Hz at a frequency offset of 1MHz from the carrier, and a floor of -168dBc/Hz starting at about 100MHz offset from the carrier

In the derivation of this model, [43] assumes that the resulting excess phase modulation φ is much smaller than 1 (small angle modulation) and that the amplitude perturbations due to noise components are negligible in comparison to the phase perturbations. Aside from these assumptions, there are two issues connected to this approach. The first is that it does not explain the up-conversion of low-frequency noise or the down-conversion of high-frequency noise to frequencies around ω_0. Indeed, Equation 4.35 contains the offset frequency ω_m and not ω, thus postulating that the flicker noise component is present at frequencies around ω_0, which is generally not true if ω_o is above some tens of MHz. The second is that F and ω_f are essentially empirical parameters, which have to be fitted to the measured phase noise profile. It also gives no indication of how to compute them based on actual circuit topology. Despite these limitations, Leeson's model correctly points out the two fundamental parameters influencing the phase noise of any oscillator the most, namely, the power consumption and the tank quality factor.

4.3.2   Oscillator Figure of Merit

It has become of widespread usage in the literature to compare different oscillators by summarizing their performance in one single number: the figure of merit (FOM). The most common and widely used FOM normalizes the phase noise number to the power consumption, the oscillation frequency, and, of course, the offset frequency at which the phase noise is measured. It is defined as:

$$\text{FOM} = \mathcal{L}(\omega_m) \left(\frac{\omega_m}{\omega_0} \right)^2 \left(\frac{P}{1\text{mW}} \right) \tag{4.38}$$

where $\mathcal{L}$ is the phase noise expressed in linear scale, P is the power of the oscillator in W, and ω_m is an offset frequency from the carrier within the $1/f^2$ phase noise region.[2] Note that the FOM is usually expressed in dB, in which case it can be written as:

$$\text{FOM} = \mathcal{L}(\omega_m) + 20\log\left(\frac{\omega_m}{\omega_0} \right) + 10\log\left(\frac{P}{1\text{mW}} \right) \tag{4.39}$$

with $\mathcal{L}$ expressed in dBc/Hz.

EXAMPLE 14   In the GSM cellular standard, the receiver specifications require a local oscillator having a phase noise better than -138dBc/Hz at 3MHz offset from a 900MHz carrier. Using the formula above, the FOM of this oscillator is equal to $-138 + 20\log(3/900) + 10\log(P/1\text{mW}) = -188 + 10\log(P/1\text{mW})$. Best in class LC-VCOs consume a few milliwatts of power. If, for instance 4mW are consumed, then FOM $= -182$dBc/Hz.

---

[2] Some authors define the same FOM to be the inverse of the expression reported here, but we will stick to the most widespread usage.

It is interesting to understand what is the real meaning of the FOM if we assume that the oscillator follows Leeson's model. By replacing Equation 4.37 limited to the $1/f^2$ region (that is removing the floor and the flicker components) in 4.38 it is straightforward to see that the FOM becomes:[3]

$$\text{FOM} = \frac{10^3 kTF}{2Q^2}. \tag{4.40}$$

Thus the FOM is basically a combined measure of the quality factor of the loaded tank and of the noise factor of the active devices. State of the art integrated LC oscillators for wireless applications display a FOM between -180dBc/Hz and -190dBc/Hz, while ring oscillators are definitely inferior in this respect, with typical FOMs ranging from -150dBc/Hz to -160dBc/Hz for well-designed VCOs. Applying Equation 4.40 to these numbers and realistically assuming $F = 2$, it turns out that the effective quality factor of state of the art LC-VCOs (in terms of Leeson's model Q) is in the range of 2 to 6, a significant degradation from the quality factor of a pure LC tank, which might well be in the range of 10 to 15. Even though the concept of tank quality factor does not apply to a ring oscillator, since there is usually no inductor or tank as such, by applying the same procedure the value of an "equivalent" effective quality factor for a ring oscillator is of the order of 0.05 to 0.2.

It has to be observed that this FOM does not consider many other aspects of an oscillator's performance as for instance tuning range, supply rejection, area, susceptibility to cross talk and others. As an example, a ring oscillator VCO can easily cover a tuning range of several octaves, while the tuning range of an integrated LC oscillator is typically limited to $+/-$ 20% of the center frequency. In choosing the right kind of oscillator and the right topology for a given application, the phase noise is just one of the many factors to consider, and sometimes is not even the most important, so the indication given by the FOM has to be taken with a grain of salt.

4.3.3   LC Oscillators

Following Leeson's approach it is relatively easy to compute an expression for the phase noise of a generic LC oscillator as shown in Figure 4.9. The noise sources are the parallel resistor R and the active transconductor G. Both of them inject a noise current into the LC tank, and, since the two are uncorrelated, the two-sided PSD of the noise current is equal to:

$$S_{i_n} = \frac{2kT}{R} + S_{i_G} \tag{4.41}$$

where S_{i_G} is the PSD of the noise current of the transconductor. It is customary to combine the two components in one single term and to write:

$$S_{i_n} = \frac{2kTF}{R} \tag{4.42}$$

where $F \geq 1$ is a noise factor considering the increase in current noise due to active elements in the loop, relative to the resistor noise alone.

---

[3] Since the power in the FOM is normalized to 1mW, a factor 10^3 needs to be introduced in the numerator.

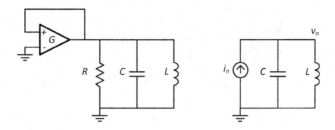

Figure 4.9 Generic LC oscillator (left) and equivalent circuit for noise calculation (right).

The PSD of the voltage noise on the tank can be calculated as $S_{v_n}(\omega) = |Z(\omega)|^2 S_{i_n}$, where $Z(\omega)$ is the impedance seen by the noise current source. At this point, it is important to note that, in steady state oscillation, the tank losses R are exactly compensated for by the transconductor G, so that the impedance seen by the noise current is that of the pure LC tank (see Figure 4.9, right):

$$Z(\omega) = \frac{j\omega L}{1 - \omega^2 LC}. \tag{4.43}$$

Considering $\omega = \omega_0 + \omega_m$, where $\omega_0 = 1/\sqrt{LC}$ is the oscillation frequency and ω_m is a small frequency offset ($\omega_m \ll \omega_0$), the impedance can be approximated as:

$$Z(\omega_0 + \omega_m) = \frac{-j}{2\omega_m C} \tag{4.44}$$

and the PSD of the output voltage noise close to the carrier can be written as:

$$S_{v_n}(\omega_0 + \omega_m) = \frac{2kTF}{R}\left(\frac{1}{2\omega_m C}\right)^2. \tag{4.45}$$

Introducing the Q factor of the tank $Q = \omega_0 RC$, the voltage noise becomes:

$$S_{v_n}(\omega_0 + \omega_m) = 2kTFR\left(\frac{\omega_0}{2\omega_m Q}\right)^2. \tag{4.46}$$

From thermodynamic considerations, the noise power is equally split into amplitude and phase of the oscillation, so that the PSD effectively contributing to phase noise is half of the previous expression. The phase noise can then finally be computed by dividing one half of the two-sided S_{v_n} by the two-sided power of the carrier $V_{tank}^2/4$, with V_{tank} the oscillation amplitude. The result is:

$$\mathcal{L}(\omega_m) = \frac{2kTFR}{V_{tank}^2/2}\left(\frac{\omega_0}{2\omega_m Q}\right)^2. \tag{4.47}$$

This result is Leeson's formula (see Equation 4.37) derived in the previous section.

Note how this approach, based on a linear time-invariant model, works as long as the only relevant noise sources are white, and fails to capture the up-conversion of low-frequency current noise or the down-conversion of high-frequency current noise into the sidebands of oscillation.

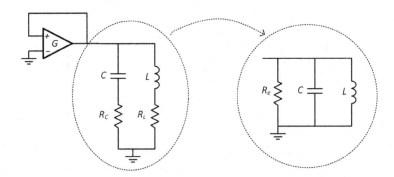

Figure 4.10 Generic LC oscillator with individual reactive element losses (left) and equivalent circuit where the losses have been lumped in one single parallel resistor (right).

The authors of [44] extended the previous approach to account for the individual losses of the reactive elements in the tank (see Figure 4.10). The current noise produced by each resistor, multiplied by its transfer function to the output voltage, are added together in an uncorrelated fashion to give the total output noise. The same result can be obtained in a much easier way by replacing each branch of a reactive element in series with a resistor by the parallel combination of the same reactive element and a proper conductance.

It can be shown that the series of a capacitor C and a resistor R_C is to a first approximation equivalent to the parallel arrangement of the capacitor C and a conductance $G_C = (\omega C)^2 R_C$. In a similar way, the series of an inductor L and a resistor R_L is to a first approximation equivalent to the parallel arrangement of L and a conductance $G_L = R_L/(\omega L)^2$. Therefore, the tank in Figure 4.10 is equivalent to the tank in Figure 4.9, provided that the parallel resistor R is replaced by an effective parallel resistor R_e:

$$\frac{1}{R_e} = G_L + G_C = \frac{C}{L}(R_L + R_C). \tag{4.48}$$

All considerations and formulas derived previously can be repeated, with R replaced by R_e. It is interesting to note how, according to this expression, to minimize the phase noise due to losses of the reactive elements of the tank for a given oscillation frequency and given R_C and R_L, the inductance has to be maximized and the capacitance minimized. Of course the values of R_C and R_L depend on C and L, too; for instance, a larger inductance needs more windings and thus produces a larger R_L, so that the trade-off is not so straightforward. Additionally, a large L might push the oscillator into the voltage-limited region of operation already at relatively low value of current [45] [46], resulting in a degradation of phase noise. This might lead to a design which is for sure very low power, but which might not be able to meet the target phase noise specifications.

4.3.4   Crystal and Other Oscillators

Crystal oscillators are circuits that exploit the piezoelectric properties of crystal quartz to generate very precise and stable clock references. The generic block diagram of a

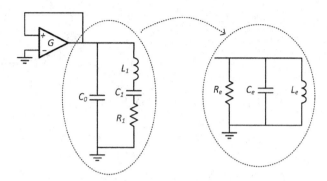

Figure 4.11 Generic crystal oscillator with crystal tank (left) and equivalent RLC circuit for the tank at parallel resonance.

crystal oscillator is shown on the left in Figure 4.11. A transconductor G pumps current into the equivalent circuit of the crystal and the voltage generated by the impedance of the tank is fed back to the input of the transconductor, closing the loop. The quartz is essentially a mechanical element, but its operation in the context of electrical oscillators can be modeled by a lumped equivalent circuit featuring a capacitor C_0 in parallel to a branch made of an inductor L_1 a capacitor C_1 and resistor R_1 in series. These elements are called the *motional* inductance, capacitance, and resistance, and model the mechanical behavior of the crystal in equivalent electrical terms. It is important to note that these elements, as well as the currents and voltages of the intermediate nodes, are nowhere to be found in the quartz; they just serve the goal of providing an equivalent electrical model. The parallel capacitor C_0, to the contrary, is a real capacitor, summing up the capacitances due to the quartz packaging and those placed on the board or on the integrated circuit to allow oscillation.

The tank shown in Figure 4.11 (left) has two resonant modes. In series resonance, the motional inductor L_1 resonates with the motional capacitance C_1 providing a low impedance path in the series branch and a phase shift from $-\pi/2$ to $+\pi/2$. In parallel resonance, L_1 resonates with the series of C_0 and C_1, providing a high impedance load to the transconductor and a phase shift back from $+\pi/2$ to $-\pi/2$. Since the phase crosses the zero twice, in both series and parallel resonance, considering the Barkhausen criterion, the circuit could oscillate at both frequencies. In the simplified configuration shown in Figure 4.11, the series resonant mode is not stable, since, from the point of view of the phase shift, it has positive feedback to the transconductor input. The parallel resonance is stable, though, and the crystal will oscillate in this mode at a frequency given by:

$$\omega_0 = \frac{1}{\sqrt{LC_{01}}} \qquad (4.49)$$

where $C_{01} = C_0 C_1/(C_0 + C_1)$ is the equivalent series capacitance of C_0 and C_1. However, in practical implementations like the very widespread Pierce configuration, the crystal is connected between the gate and the drain of an active element and usually additional load capacitors are inserted between each of the two crystal terminals and

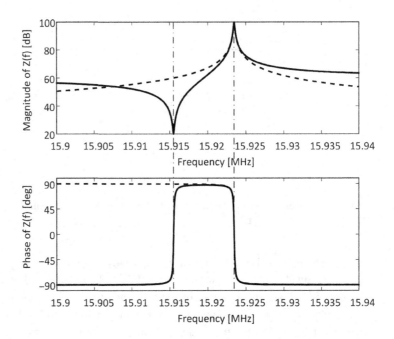

Figure 4.12 Magnitude and phase plot of the impedance of a crystal tank (solid) and approximation (dashed) at parallel resonance.

ground. Also, efforts are made to reduce any capacitance in parallel to the crystal, which would make the oscillation condition harder. In this configuration, the crystal oscillates at a frequency between the series and the parallel resonant frequencies, where its behavior is inductive. The larger the values of the load capacitors, the closer the oscillation frequency will be to the series resonance. For an overview of the physical properties of quartz crystals and a complete analysis of oscillator circuits the reader is referred to [47], [48], and [34].

It is instructive to get a feeling for the values of the electrical components involved in a fundamental tone crystal. Typically, C_0 is in the range of few pF. C_1, to the contrary, is very small, in the range of some fF, about one thousandth of C_0. The motional inductor L_1 is huge, in the range of few mH (though this is only a model; there is no such inductor in the quartz), and the motional resistance R_1, modeling the losses in the tank, is a few Ohms. Figure 4.12 shows the magnitude and phase plots of the impedance seen by the transconductor for typical values of $C_0 = 10$pF, $C_1 = 10$fF, $L_1 = 10$mH, and $R_1 = 10$Ohm. The two resonant frequencies are clearly visible and are quite close to each other (due to the fact that C_1 is very small compared to C_0).

In order to understand the phase noise of a crystal oscillator and reuse the results derived before in the case of the LC tank, we seek to approximate the behavior of the impedance at resonance with a parallel RLC network. The admittance of the crystal tank is given by:

$$Y(s) = \frac{s[C_0 + C_1 + sC_0C_1(R_1 + sL_1)]}{1 + sC_1(R_1 + sL_1)}. \tag{4.50}$$

For frequencies around resonance, the denominator can be approximated as:

$$1 + sC_1(R_1 + sL_1) \approx \frac{C_1}{C_0 + C_1} s^2 C_1 L. \tag{4.51}$$

Replacing this expression in Equation 4.50 and after some easy algebra, the tank admittance can be written as:

$$Y(s) = \left(\frac{C_0 + C_1}{C_1} \right)^2 \left(\frac{1}{sL_1} + \frac{R_1 C_{01}}{L_1} + sC_{01} \right) \tag{4.52}$$

which turns out to be the parallel of an equivalent inductor L_e, capacitor C_e, and resistor R_e according to:

$$L_e = L_1 \left(\frac{C_1}{C_0 + C_1} \right)^2 \tag{4.53}$$

$$C_e = C_{01} \left(\frac{C_0 + C_1}{C_1} \right)^2 \tag{4.54}$$

$$R_e = R_1 Q^2 \left(\frac{C_1}{C_0 + C_1} \right)^2 \tag{4.55}$$

with $Q = 1/(R_1 \omega_0 C_{01})$ the quality factor of the tank. The equivalent circuit is shown on the right in Figure 4.11 and the corresponding Bode diagram is shown in Figure 4.12 with dashed line. At this point, it is straightforward to derive an expression for the phase noise of the oscillator, following the very same steps as in Section 4.3.3. The result is exactly equal to Equation 4.47, where V_{tank} is the oscillation amplitude across the equivalent resistor R_e.

The difference between a crystal oscillator and an integrated LC oscillator, from the phase noise point of view, is exclusively in the value of the quality factor Q. While integrated LC tanks have quality factors typically ranging from 5 to 20, mainly limited by the losses in the substrate and series resistance of the windings, a crystal has a Q in the range of thousands to a few millions. A typical value of $Q = 10^5$ leads to the plot in Figure 4.12. Another advantage of the crystal is that, due to the extremely low losses, it is very suitable for ultra-low power clock generation. On the other hand, due to fabrication and robustness issues, the sizes of the quartz crystal cannot be made too small; thus, the devices are bulky, cannot be integrated on silicon, and have frequency ranges limited to about 50MHz. Overtone crystals can work at frequencies above 100MHz but they are quite delicate and costly, so that they are not used in typical high-volume commercial applications.

The framework outlined in this section can be used to investigate not only the phase noise of crystal oscillators, but also of all those oscillators which can be modeled by a lumped electrical network as shown in Figure 4.11 (left), such as, for instance, MEMS, SAW, and BAW oscillators.

Note that in all high-Q oscillators, the phase noise performance is typically limited by the noise produced by the sustaining circuitry, rather than by the intrinsic losses of the tank. In this sense the factor F in Equation 4.47 plays a very fundamental role and can be quite large, depending on the specific circuit implementation.

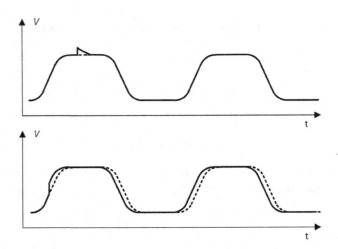

Figure 4.13 Simplified waveforms showing how noise injected in the same node at different times, for instance in a CMOS inverter-based ring oscillator, have different effects on the output jitter.

4.4   Linear Time-Variant Analysis

The analysis of jitter and phase noise carried out in the previous sections treats the circuits under investigation as linear time-invariant systems, ignoring the fact that every oscillator is actually a periodic time-variant system.

Indeed, noise injected into a given node of the oscillator at different times might affect the output jitter to different degrees. For example, take a simple CMOS inverter-based ring oscillator: the noise current injected by a MOS device when the respective output node is saturated to ground, or V_{DD}, will be "absorbed" by the power and ground supplies and will not significantly change the following shape of the voltage waveform. Therefore its effect on the output jitter will be minimal. On the other hand, the same amount of noise injected in the same node during a charging or discharging phase will definitely impact the voltage waveform, and will produce jitter at the output (see Figure 4.13).

Based on this observation, Hajimiri and Lee developed a theory of noise in oscillators, which takes into consideration the time-varying nature of the response of the circuit to noise injected into its nodes [49], [50], [51], [52]. They introduced the concept of the *impulse sensitivity function (ISF)*, a periodic function which describes the relationship between an impulse of noise current injected at a given time on a given node, and the jitter at the output of the oscillator. By elaborating further on this concept, they set a framework for the time variant analysis of any kind of oscillator, which led to numerous valuable insights into the conversion of amplitude noise into jitter.

4.4.1   The Impulse Sensitivity Function

Although a detailed analysis of ISF theory is not the focus here, it is instructive to understand the mechanism behind it. A small packet of charge injected into a specific node of the circuit at time τ is assumed to affect the output absolute jitter $\mathbf{a}(t)$ over time t in a proportional way. The proportionality is given by a function $h_{\mathbf{j}}(t, \tau)$ defined as:

$$h_{\mathbf{j}}(t, \tau) = \frac{-\Gamma(\tau)}{\omega_0 q_{max}} u(t - \tau) \qquad (4.56)$$

where ω_0 is the oscillation frequency, q_{max} is the peak amount of charge due to the oscillation on the capacitance of the node, $u(t)$ is the unity step function and $\Gamma(\tau)$ is the ISF. In terms of excess phase, the defining equation of the ISF is obtained by using the relation in Equation 3.5, and assumes the better known form:

$$h_\varphi(t, \tau) = \frac{\Gamma(\tau)}{q_{max}} u(t - \tau). \qquad (4.57)$$

The ISF depends, of course, on the circuit topology, but investigations show that it has a strong correlation with the shape of the oscillation waveform itself. The total output jitter due to a current noise $i_n(\tau)$, injected into the specific node, is given by the sum of the small noise charge packets injected at each time instant τ, multiplied by $h_{\mathbf{j}}(t, \tau)$. Expressing this as an integral:

$$\mathbf{a}(t) = \int_{-\infty}^{+\infty} h_{\mathbf{j}}(t, \tau) i_n(\tau) d\tau = \frac{-1}{\omega_0 q_{max}} \int_{-\infty}^{t} \Gamma(\tau) i_n(\tau) d\tau \qquad (4.58)$$

or, in terms of excess phase:

$$\varphi(t) = \int_{-\infty}^{+\infty} h_\varphi(t, \tau) i_n(\tau) d\tau = \frac{1}{q_{max}} \int_{-\infty}^{t} \Gamma(\tau) i_n(\tau) d\tau. \qquad (4.59)$$

The first key observation is that, due to the periodic nature of the oscillation, the same amount of charge injected into the same node at two time instants one or more oscillation periods apart must have the same effect on the output jitter. Therefore, the ISF function is periodic in ω_0 and can thus be expressed in a Fourier series as:

$$\Gamma(\tau) = \frac{c_0}{2} + \sum_{n=1}^{+\infty} c_n \cos(n\omega_0 \tau + \theta_n). \qquad (4.60)$$

The second key observation is that we are interested in behavior of $\mathbf{a}(t)$ (or of $\varphi(t)$) versus time t at frequencies much lower than the oscillation period. Higher frequency components are not significant here since they are averaged out over a number of periods of the oscillation. Due to the integrals in Equations 4.58 and 4.59, which are in effect a mixing product, and considering Equation 4.60, only those noise frequency components which are close to zero or to multiples of ω_0 will have a significant contribution to the jitter or excess phase at frequencies close to zero. The others are averaged out by the integral.

To clarify this point, we follow the approach in [49]. Considering only the n-th harmonic component of the ISF function $c_n \cos(n\omega_0 \tau + \theta_n)$ and assuming the noise current to be located at an offset frequency ω_m from $n\omega_0$, expressed as $i_n(\tau) = i_n \cos((n\omega_0 + \omega_m)\tau)$, the integral 4.59 can be written as:

$$\varphi(t) = \frac{c_n i_n}{2 q_{max}} \int_{-\infty}^{t} (\cos(2n\omega_0 \tau + \omega_m \tau + \theta_n) + \cos(\omega_m \tau - \theta_n)) d\tau \qquad (4.61)$$

where the trigonometric formula for the multiplication of two cosines has been used. The first term in the integral results in an excess phase component at double the oscillation frequency and can be dropped, since we are interested in excess phase variations at frequencies much lower than the carrier. The second term, on the contrary, is at low frequency and is relevant for the computation of the phase noise. Working out the integral, the result is:

$$\varphi(t) = \frac{c_n i_n}{2 q_{max} \omega_m} \sin(\omega_m t - \theta_n). \tag{4.62}$$

From this expression, it can be understood that the excess phase PSD at frequency ω_m is proportional to the superposition of the noise current PSD at frequencies $n\omega_0 + \omega_m$, for each n, multiplied by the corresponding squared Fourier coefficient of the ISF function c_n^2. Note also that another effect of the integral in Equation 4.59 is that of generating a factor ω_m, the offset of the noise component from the nearest multiple of the oscillation frequency, in the denominator of the result. For this reason, a white (flat) current noise PSD is converted into a $1/f^2$ PSD at low frequencies and flicker noise PSD is converted into a $1/f^3$ PSD This mechanism is depicted in Figure 4.14.

One notable result of this analysis is that low-frequency noise, in particular flicker noise, is converted into jitter by the multiplication of its PSD with the coefficient c_0^2. Since c_0 is the DC (average) value of the ISF, according to this theory, the conversion of flicker noise into jitter can be greatly reduced or even eliminated by making the oscillation more symmetric, such that the DC value of the ISF is zero.[4] From Figure 4.14 it can also be noted that, if c_0 is small enough, the $1/f^3$ corner of the jitter PSD can be significantly lower than the $1/f$ corner of the device noise.

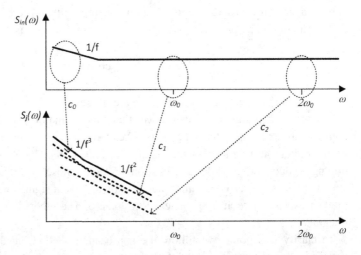

Figure 4.14 Conversion of current noise into low-frequency jitter according to the ISF theory. In the bottom plot, the thin lines are the individual contributions, while the thick line is the resulting total.

---

[4] Note that, strictly speaking, the 1/f noise waveform, assumed to be cyclostationary, must be symmetrical too with respect to the ISF.

If the current noise source is white and stationary, then the resulting PSD is proportional to the sum of all Fourier coefficients squared. By virtue of Parseval's theorem, that value is also equal to the square of the RMS value of the ISF. A detailed analysis shows that under this conditions the phase noise can be expressed as [49], [50]:

$$\mathcal{L}(\omega_m) = \frac{\Gamma_{rms}^2 S_i}{(\omega_m q_{max})^2} \tag{4.63}$$

where S_i is the two-sided PSD of the current noise sources (assumed to be white), and Γ_{rms}^2 is the RMS value of the ISF:

$$\Gamma_{rms}^2 := \frac{1}{T_0} \int_0^{T_0} \Gamma^2(\tau) \, d\tau = \frac{1}{2} \left(\frac{c_0^2}{2} + \sum_{n=1}^{\infty} c_n^2 \right) \tag{4.64}$$

with $T_0 = 2\pi/\omega_0$ the period of the oscillation. Thus, in order to minimize jitter or excess phase due to white noise sources, the RMS value of the ISF should also be minimized. Note that in the original paper [50], the equation corresponding to Equation 4.63 has an additional factor 2 in the denominator, since there the *one-sided* current noise PSD S_i' was considered instead of the *two-sided*.[5]

In many practical cases the current noise sources are not stationary, though. The noise produced by active devices depends on the biasing point of the device itself, which, in an oscillator, is typically not constant; rather, it is periodic with the same period as the oscillation. The noise sources are thus essentially cyclostationary in nature. Equation 4.63 can be extended to the case of a cyclostationary PSD $S_i(\tau)$ by observing that the cyclostationary noise source can be modeled as a stationary noise source multiplied by a periodic modulation function. The *effective* ISF is then the product of the original ISF and the modulating function itself [49]. With easy mathematical derivation, replacing the ISF in Equation 4.63 with the effective ISF, the phase noise in presence of cyclostationary noise sources can be written as:

$$\mathcal{L}(\omega_m) = \frac{\overline{\Gamma^2(\tau) S_i(\tau)}}{(\omega_m q_{max})^2} \tag{4.65}$$

where:

$$\overline{\Gamma^2(\tau) S_i(\tau)} := \frac{1}{T_0} \int_0^{T_0} \Gamma^2(\tau) S_i(\tau) \, d\tau. \tag{4.66}$$

4.4.2 Application to Ring Oscillators

In [51] this theory is applied to the case of inverter-based and differential stage ring oscillators. The result for the phase noise of an inverter based ring is:

$$\mathcal{L}(f) = \frac{8}{3\eta} \frac{kT}{P} \frac{\gamma V_{DD}}{\Delta V} \frac{f_0^2}{f^2} \tag{4.67}$$

---

[5] For a detailed discussion on one-sided versus two-sided PSD, refer to Appendix A.2.8.

where P is the total power consumption and η is a "form" factor, typically close to 1, which depends on the shape of the oscillation waveforms. The parameter ΔV is a characteristic overdrive voltage, defined as $\Delta V = V_{DD}/2 - V_{th}$ if the transistor is in quadratic current operation region (long channel devices), and as $\Delta V = E_c L$ for transistors in velocity saturation (a possible condition for short channel devices), with E_c the critical electric field. For a differential stage ring, the work in [51] neglects the noise coming from the tail current and obtains the result:

$$\mathcal{L}(f) = \frac{8}{3\eta}kT\frac{M}{I_{tot}V_{op}}\left[1 + \frac{\gamma V_{op}}{(\Delta V)_d}\right]\frac{f_0^2}{f^2} \tag{4.68}$$

where in this case ΔV is equal to $V_{GS} - V_{th}$ for quadratic current operation region, and to $E_c L$ for velocity saturation. By using Equation 3.56 the period jitter can be calculated to be:

$$\sigma_{\mathbf{p}}^2 = \frac{8}{3\eta}\frac{kT}{f_0 P}\frac{\gamma V_{DD}}{\Delta V} \tag{4.69}$$

for the inverter-based ring, and:

$$\sigma_{\mathbf{p}}^2 = \frac{8}{3\eta}kT\frac{M}{f_0 I_{tot}V_{op}}\left[1 + \frac{\gamma V_{op}}{(\Delta V)_d}\right] \tag{4.70}$$

for the differential ring. These expressions may be compared to Equations 4.22 and 4.23 obtained with a more classical analysis. The main differences among these equations, apart from the presence of the factor η, is to be found in the expression for the inverter-based ring for quadratic current regime. While the classical analysis leads to a variance inversely proportional to $V_{DD} - V_{th}$, in the time variant one the variance is inversely proportional to $V_{DD}/2 - V_{th}$. The reason for this discrepancy is that [51] considers a finite transition time of the input slope, thus identifying the overdrive voltage of the inverter as $V_{DD}/2 - V_{th}$, while [37] and [33] assume an infinite input slope, leading to an overdrive voltage of $V_{DD} - V_{th}$. For advanced CMOS technologies, where the ratio of supply voltage to threshold voltage is lower, and not minimum length devices, the two expressions might lead to significantly different results.

4.4.3   The ISF of an LC Tank

To apply the linear time-variant analysis to LC oscillators, it is necessary first to derive the expression for the ISF of an LC tank. With reference to Figure 4.15, let's assume that a lossless LC tank is sustaining an oscillation at amplitude V_{tank} and frequency ω_0 as in:

$$v_0(t) = V_{tank}\sin(\omega_0 t). \tag{4.71}$$

To derive the ISF, we inject a small packet of charge Δq at time τ into the tank, and find out what is the resultant change in the timing of the rising edge of the oscillation. Since the LC is a linear system, its total response is the sum of the existing oscillation (Equation 4.71) and the oscillation that would be generated by the charge injection at

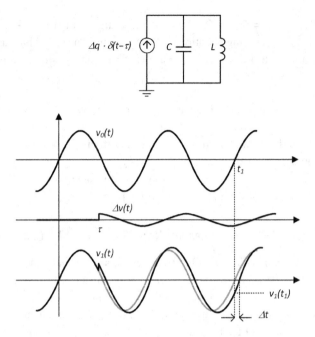

Figure 4.15 Injection of a charge packet into an ideal LC tank.

zero initial conditions. At zero initial conditions, injecting a charge Δq in the tank results in voltage step on the capacitor equal to $\Delta q / C$, which triggers an oscillation $\Delta v(t)$ at frequency ω_0 with maximum voltage at time τ. From these considerations, the total response of the LC tank can be written as:

$$v_1(t) = V_{tank} \sin(\omega_0 t) + \frac{\Delta q}{C} \cos(\omega_0 t - \omega_0 \tau). \tag{4.72}$$

The rising edges of the unperturbed oscillation occur at time t_1 such that $\sin(\omega_0 t_1) = 0$ and $\cos(\omega_0 t_1) = 1$. By replacing these conditions in 4.72, where the cosine term has been expanded by using the usual trigonometric formulas for the sum and difference of angles, the amplitude of the perturbed oscillation at t_1 is found to be:

$$v_1(t_1) = \frac{\Delta q}{C} \cos(\omega_0 \tau). \tag{4.73}$$

Assuming a small perturbation ($\Delta q / C \ll V_{tank}$), the change in the rising edge timing can be derived by dividing the amplitude value at t_1 by the slope of the signal at t_1:

$$\Delta t \approx - \left(v_1(t) \bigg/ \frac{\partial v_0}{\partial t} \right) \bigg|_{t=t_1} = -\frac{\Delta q}{C V_{tank} \omega_0} \cos(\omega_0 \tau) \tag{4.74}$$

and finally, converting time difference into phase difference using Equation 3.5, the change in the phase of the signal is:

$$\Delta \varphi = -\Delta t \cdot \omega_0 = \frac{\Delta q}{q_{max}} \cos(\omega_0 \tau) \tag{4.75}$$

where $q_{max} = CV_{tank}$ is the maximum charge in the LC tank. By comparing this expression to the defining equation of the ISF, Equation 4.57, which assumes a current impulse of amplitude 1 and thus $\Delta q = 1$, the ISF for an LC tank assumes the very simple form:

$$\Gamma(\tau) = \cos(\omega_0\tau). \qquad (4.76)$$

The ISF is in quadrature with the oscillation waveform: noise injected when the signal has its maximum amplitude has no impact on the excess phase, while noise injected around the zero amplitude has a maximum effect. Note also that the ISF is simply the time derivative of the original oscillation. This is not a mere coincidence, since it has been proven that the ISF is strictly linked to the derivative of the oscillation voltage for a wide class of oscillating systems [49].

4.4.4 A General Result on Leeson's Noise Factor F for LC Oscillators

The analysis of LC oscillators in Sections 4.3.3 and following was carried out assuming a linear time-invariant system, but the result in Equation 4.76 proves that the noise injected in an LC tank has a time-dependent effect on the excess phase of the system, therefore a time-variant approach is needed. The application of the ISF theory to LC-based oscillator leads, under very general assumptions, to the surprising and remarkable result that the noise factor F is largely independent of the specific nature and operation conditions of the active device, independent of the shape of the current, and depends mainly on the tank losses and the topology of the oscillator. In this section we follow the approach used by [53], leveraging the ISF theory. The same result, although with different notations, is obtained also by [54], with a more general approach using descriptive functions, and by [55], extending the phasor-based analysis pioneered by [56].

Equating the general Equation 4.65 for phase noise obtained by using the ISF approach to Leeson's descriptive formula in Equation 4.37, limited to the $1/f^2$ portion:

$$\frac{\overline{\Gamma^2(t)S_i(t)}}{(\omega_m q_{max})^2} = \frac{2kTF}{P}\left(\frac{\omega_0}{2Q\omega_m}\right)^2 \qquad (4.77)$$

where $q_{max} = CV_{tank}$, V_{tank} is the voltage across the tank, $P = V_{tank}^2/(2R)$, $Q = \omega_0 CR$, and R is the total loss in parallel to the tank, the following expression for the noise factor F as a function of the ISF can be found:

$$F = \frac{R}{kT} \cdot \overline{\Gamma^2(t)S_i(t)}. \qquad (4.78)$$

In the following, we will apply Equation 4.78 to calculate the factor F for the general LC oscillator shown in Figure 4.16, considering noise generated by the tank losses and by the transconductor active devices. Note that in this generic topology, it is not assumed that the voltage across the tank is equal to the voltage either at the input of the transconductor or at its output. It has indeed been amply well shown in the literature that in the most efficient high-purity LC oscillators, these quantities are generally different. For instance, the current generated by the transconductor can be injected into only part of the tank, and the tank voltage can in general be transformed by means of a lossless network before being fed back to the transconductor.

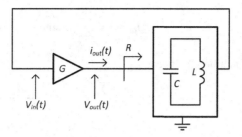

Figure 4.16 Generic LC harmonic oscillator.

The corresponding two-sided PSD of the current noise is given by:

$$S_{i,R} = \frac{2kT}{R} \qquad (4.79)$$

and applying Equation 4.66 it can be easily shown that:

$$\overline{\Gamma^2(t)S_{i,R}} = \frac{1}{T_0} \int_0^{T_0} \cos(\omega_0 t)^2 \frac{2kT}{R} \, dt = \frac{kT}{R}. \qquad (4.80)$$

Note that this expression, combined with Equation 4.78, returns $F = 1$, which is exactly the result obtained in the time-invariant approach when the noise of the transconductor is neglected. With this approach, though, there is no need to invoke thermodynamics and the equal splitting of the power in amplitude noise and phase noise. The factor 1/2 descends naturally from the form of the ISF. An intellectually very pleasing result.

In evaluating the effect of the noise of the transconductor, it will be assumed that the oscillation waveform is sinusoidal, that the noise of the transconductor is proportional to the transconductance of the active devices, and that the transconductor devices operate as transistors. The first assumption generally holds if the oscillator is operated in the current limited regime [45], and the second and third are true if the active devices are operated in saturation. Both of these assumptions are very reasonable, since they both lead to a maximally efficient LC design in terms of better phase noise and lower power. With reference to Figure 4.16, the PSD of the current noise generated by the transconductor is:

$$S_{i,G}(t) = 2kT\gamma g_m(t) \qquad (4.81)$$

where the transconductance $g_m(t)$ can be expressed as function of the input voltage $v_{in(t)}$ and output current i_{out} as:

$$g_m(t) = \frac{\partial i_{out}}{\partial v_{in}} = \frac{\partial i_{out}}{\partial t} \bigg/ \frac{\partial v_{in}}{\partial t}. \qquad (4.82)$$

The input voltage is assumed to be sinusoidal:

$$v_{in}(t) = V_{in} \sin(\omega_0 t) \qquad (4.83)$$

while the output current, periodic in T_0, is expressed without loss of generality as Fourier series:

$$i_{out}(t) = \sum_{n=0}^{\infty} I_n \sin(n\omega_o t + \phi_n) \tag{4.84}$$

so that Equation 4.82 becomes:

$$g_m(t) = \frac{\sum_{n=1}^{\infty} nI_n \cos(n\omega_o t + \phi_n)}{V_{in} \cos(\omega_0 t)}. \tag{4.85}$$

As can be seen, the transconductance is periodic in T_0 and the noise generated is cyclostationary. Applying Equation 4.66 we obtain:

$$\overline{\Gamma^2(t)S_{i,G}(t)} = \frac{2kT\gamma}{V_{in}} \sum_{n=1}^{\infty} \left[\frac{nI_n}{T_0} \int_0^{T_0} \cos(\omega_0 t)\cos(n\omega_o t + \phi_n)\, dt \right]. \tag{4.86}$$

The integral above is nonzero only for $n = 1$, and considering that $\phi_1 = 0$ since at resonance the fundamental harmonic of the current is in phase with the voltage, the expression can be simplified to:

$$\overline{\Gamma^2(t)S_{i,G}(t)} = \frac{2kT\gamma}{V_{in}} \cdot \frac{I_1}{2} = \frac{kT\gamma}{R} \cdot \frac{V_{out}}{V_{in}} \tag{4.87}$$

where I_1 is the fundamental harmonic of the current and:

$$V_{out} = RI_1 \tag{4.88}$$

is the amplitude of the voltage at the output of the transconductor, considering that all other harmonics are eliminated by the LC selective filtering.

Combining Equations 4.78, 4.79, and 4.87, finally the following surprisingly simple expression can be obtained for the noise factor F:

$$F = 1 + \gamma \frac{V_{out}}{V_{in}}. \tag{4.89}$$

As stated before, this remarkable result shows that the noise factor only depends on the ratio of output voltage to input voltage of the transconductor and does not depend on any specific property of the transconductor itself, except for γ. Note that the application of the time-invariant approach to the noise of the transconductor would have led to a noise factor F erroneously dependent on g_m. It has to be understood that the ratio V_{out}/V_{in} is greatly determined by the topology of the oscillator, and has little or no dependency on other parameters like, e.g., the nature of the transconductance element or the shape of the current. Referring to Leeson's formula, Equation 4.37, it can be understood that the biggest lever to obtain lower phase noise at lower power is in the right choice of the oscillator topology, aside from the obvious criteria of reducing the tank losses and increasing the oscillation amplitude as much as possible. This aspect will be illustrated in the next section.

4.4.5   Application to Some Common LC Oscillator Topologies

In this section we will apply the general result obtained in Section 4.4.4 to some of the most common LC oscillator topologies, namely the single-switch (SS) cross coupled,

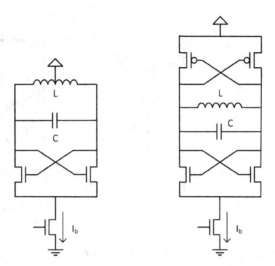

Figure 4.17 Common Single-Switch (left) and Double-Switch (right) LC oscillator topologies.

the double-switch (DS) cross coupled and the Colpitts. Additionally we will apply it also to the Class-C LC oscillator, a topology which has been shown able to achieve excellent phase noise performance at low power (see [53]).

Figure 4.17 shows the topologies for the classical SS and the DS LC oscillators. From an inspection of the figure it is immediate to understand that the voltage at the output of the transconductance element is equal to the voltage at its input and is equal also to the voltage across the tank. The fact that one topology has only one cross-coupled pair and the other has two does not change the fact that $V_{in} = V_{out}$ for both of them. Therefore, the noise factor for these topologies is the same and equal to:

$$F_{SS} = F_{DS} = 1 + \gamma. \tag{4.90}$$

This result tells us that the four transistors in the DS topology produce the same effective noise as the two transistors in the SS topology. The reason for that is that in the DS topology the switching of the cross-coupled pair is faster than in the SS, therefore the fraction of time when the devices produce effective noise is less in the DS then in the SS [57]. It has to be noted, though, that this does not mean that the two topologies have the same phase noise. Indeed the amplitude of oscillation V_{tank} is different in the two cases. In the SS, the tail bias current flows into only half of the tank, due to the supply tap in the middle of the inductor, while in the DS, because of the complementary PMOS switches at the top, it flows into the whole tank. The oscillation amplitude is therefore double in the DS than it is in the SS topology. A precise analysis shows that:

$$V_{tank,SS} = \frac{2}{\pi} I_b R \tag{4.91}$$

and

$$V_{tank,DS} = \frac{4}{\pi} I_b R. \tag{4.92}$$

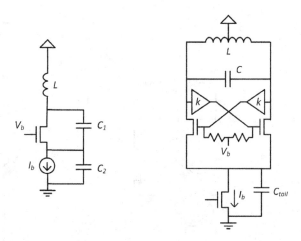

Figure 4.18 Colpitts (left) and Class-C (right) LC oscillator topologies.

Since the tank voltage is double in the DS compared with the SS, the DS topology shows 6dB better phase noise than the SS, when the oscillators are operated in the current-limited region. This result has been proven experimentally in [57]. Note, however, that the supply voltage of the SS switch can in general be half of that of the DS, so the efficiency of the two oscillators in term of phase noise over power is the same. A more detailed analysis, which is beyond the scope of this book, shows that the parasitic caps to ground in the two branches of the oscillator tilt the scale again in favor of the SS topology. The interested reader is referred to [57] and [55] for more details.

Figure 4.18 shows on the left the classical topology of a Colpitts oscillator. Considering that for this topology V_{in} is the gate-source voltage of the transistor, and V_{out} is the drain-source voltage, an elementary analysis of the capacitive divider C_1 and C_2 leads to the result that $V_{out}/V_{in} = C_2/C_1$, so that:

$$F_{Colpitts} = 1 + \gamma \frac{C_2}{C_1}. \tag{4.93}$$

A thorough analysis shows that the voltage across the tank is given by:

$$V_{tank,Colpitts} \approx 2I_b R \frac{C_2}{C_1 + C_2} \tag{4.94}$$

so that the resulting phase noise depends on the ratio of the capacitors C_1 and C_2: a large C_2 improves the output voltage, but it increases the noise factor, while a small C_2 is good for the noise factor, but is bad for the oscillation amplitude. It can be shown that for values of $\gamma \approx 2/3$ the ratio C_2/C_1 giving the lowest phase noise is very close to 2. However, also with this optimum setting, it turns out that the phase noise of a Colpitts oscillator is slightly worse than that of a cross-coupled LC oscillator. A complete analysis of this oscillator can be found in [58].

Figure 4.18 shows on the right a differential Class-C oscillator [53]. This topology can be thought of as being derived from a differential Colpitts oscillator by reducing the capacitor C_1 to zero and inserting a feedback network from the output to the input of the transconductor. The characteristics of this topology are the capacitance in parallel to the tail current source, the fact that the active devices are biased to be always in saturation, and that the output voltage is fed back to the input of the transconductors by means of a lossless passive network with gain k. This network can be a simple capacitive coupling, in which case $k \approx 1$ or a transformer-based feedback, in which case k can also be larger than one. Note that in this last case the biasing of the devices can be provided using the center tap of the secondary circuit of the transformer. By inspection of the topology it is immediate to see that:

$$F_{ClassC} = 1 + \frac{\gamma}{k} \qquad (4.95)$$

therefore, by making $k > 1$, we can obtain an advantage in term of noise factor with respect to the traditional SS and DS cross coupled topologies. Besides that, the key aspect of this topology is the high efficiency in converting bias current into oscillation amplitude. The combination of the large tail capacitance[6] and of the devices in saturation for all the period of the oscillation, leads to a class-C operation, with narrow and large pulses of current injected into the tank. The resulting tank voltage is:

$$V_{tank,ClassC} = I_b R \qquad (4.96)$$

which is higher than that of the SS cross-coupled, even if both feature an inductor with center tap connected to the supply. The combination of these factors makes in principle the Class-C differential oscillator a very efficient topology in terms of phase noise over power consumption. However note that in Class-C, the active transistor must be kept in saturation. This may eat up a non-negligible fraction of the supply, therefore reducing the oscillation amplitude and thus the advantages compared to other topologies. Other CMOS oscillators that show very high efficiency can be found in [59], [60], [61], [62], and [63].

4.5   Comparison of Best Achievable FOM

In this section we will compare the FOM for the oscillator topologies previously described in this chapter. The intent is to investigate the best FOM theoretically achievable for ring, relaxation, and LC oscillators, assuming realistic conditions for the technology and design parameters. The result will give the reader useful insight into the performance of each topology, and a metric to evaluate the quality of any particular design.

As a starting point we derive an equation relating the period jitter to the FOM, assuming that the oscillator is dominated by white thermal noise and thus has a $1/f^2$ phase noise profile. Combining Equations 3.56 and 4.38, it is easy to find that the FOM can be expressed as a function of the RMS period jitter $\sigma_\mathbf{p}$, oscillation frequency f_0 and power consumption P as:

$$\mathrm{FOM} = 10^3 f_0 P \sigma_\mathbf{p}^2 \qquad (4.97)$$

---

[6] The reader should be warned though that there is an upper limit to the size of this capacitor, to avoid the instability phenomenon known as *squegging* [53].

where the factor 10^3 is due to the fact that the power P in the FOM is expressed in milliwatts, while in the formula above it is considered expressed in watts.

For the case of a inverter-based ring oscillator, using Equation 4.22 obtained in the Linear Time-Invariant analysis, the best FOM achievable at room temperature ($T = 300K$) is:

$$\text{FOM} = 10^3 \cdot 2kT \cdot (1 + 4\gamma) = -165\text{dB} \tag{4.98}$$

where it has been assumed that $V_{DD} - V_{th} = V_{DD}/2$. The factor γ is set equal to its classical value 2/3 in this whole section. This result agrees with that published in [38]. It is interesting to note that, if the Linear Time-Variant analysis result of Equation 4.69 is used, realistically assuming $\eta = 1$ and $\Delta V = V_{DD}/4$, we obtain FOM $= 10^3 \cdot 2kT \cdot 5.3\gamma$, yielding the same numerical result of -165dB.

For the case of a differential stage based ring oscillator, using Equation 4.23, the FOM is:

$$\text{FOM} = 10^3 \cdot 2kT \cdot (1 + 3.5\gamma) \cdot 1.9M = -160\text{dB} \tag{4.99}$$

with the assumptions that the overdrive $V_{GS} - V_{th} = V_{DD}/4$, the number of stages is the minimum for a differential structure $M = 2$, and the voltage swing is $V_{op} = 3V_{DD}/4$, already an extreme value for a differential stage which could be obtained with nonlinear loads. Comparing the two previous equations, it is evident how the differential stage ring oscillator is intrinsically inferior to the inverter-based one, and the difference in FOM is directly related to the number of stages M.

For the relaxation oscillator topology analyzed in Figure 4.6, the RMS period jitter is given by Equation 4.30. The best achievable noise performance can be derived by making the reference voltage capacitors C_L and C_H very large, so that the noise contribution for the reference voltage is negligible compared to the noise coming from the current source. The minimum power consumption $P = C\Delta V f_0 V_{DD}$ is given by the amount of charge needed to charge the capacitor C between the two threshold levels V_L and V_H, given by $C\Delta V$, divided by the oscillator period $1/f_0$ to obtain the average current, and finally multiplied by the supply voltage V_{DD}. With this information, combining Equations 4.30 and 4.97, and again assuming $V_{GS} - V_{th} = V_{DD}/4$, the best achievable FOM results in:

$$\text{FOM} = 10^3 \cdot 2kT \cdot 4\gamma = -166\text{dB}. \tag{4.100}$$

Interestingly, this number is essentially the same as the one obtained for inverter based ring oscillators. Indeed the two oscillators share the same basic principle, which is a MOS device charging a capacitor with a constant current. The authors in [38] analyze a relaxation oscillator topology where the charging and discharging of the capacitor is obtained via a resistor alternatively connected to supply and to ground, and arrive at a FOM which is about 3dB better than this one. However, note that both analyses do not consider other important noise sources, like the bias circuits needed to generate the currents and the noise of comparators and latches. In practice, relaxation oscillators are normally used as auxiliary clock sources. As such they have to be cheap, both in terms of area and in terms of power, so that often their FOM is much worse than the limit reported here.

For LC oscillators we will use Equation 4.40, which expresses the FOM as a function of the noise factor F and of the quality factor Q. The analysis of Section 4.4.4 states that under reasonable assumptions the noise factor is largely dominated by the circuit topology, and is in expressed by Equation 4.89. Using the widespread Single-Switch or Double-Switch topologies as a reference, where the quality factor $F = 1 + \gamma$, the FOM can be expressed as:

$$
\text{FOM} = 10^3 \cdot 2kT \cdot (1 + \gamma) \cdot \frac{1}{(2Q)^2} =
\begin{cases}
-175\text{dB} & \text{if } Q = 1 \\
-189\text{dB} & \text{if } Q = 5 \\
-195\text{dB} & \text{if } Q = 10 \\
-198\text{dB} & \text{if } Q = 15 \\
-201\text{dB} & \text{if } Q = 20.
\end{cases}
\tag{4.101}
$$

From this expression it is evident how the advantage of an LC oscillator with respect to a ring or relaxation oscillator is given by the quality factor of the tank. The difference in FOM, depending on the specific value of Q, can easily reach 30dB. At a very high level, this should not be surprising. In a ring or relaxation oscillator, the current used to charge the capacitors of the oscillating nodes is directly taken from the supply, and subsequently dumped to ground during the discharge phase. So essentially the current is used only once per oscillation period. In a LC with reasonably high Q, the current is constantly exchanged between the inductor and the capacitor, and only a small part is lost in thermal dissipation and needs to be replaced by the supply. The current is thus reused over multiple oscillation periods, hence the dramatic difference in energy efficiency for the same noise performance.

Figure 4.19 offers an overview of the FOM achieved by oscillators published in the last two decades in main international journals and conferences. It shows that the limits calculated above, shown as dashed lines, hold true. In particular, notice how the FOM of relaxation oscillators is still very far from the theoretical limit, and how the performance of ring oscillators is hard limited around -160dB. Also it is evident that LC oscillators outperform any other category. The best in class oscillators, with respect to FOM, are indicated in black, and refer to the publications [53], [64], [65], [66], [67], [68], [69], and [70].

4.6   A Note on Flicker Noise

The reader has probably noticed that in the previous sections we were deriving formulas for jitter or phase noise in the presence of white noise sources in the circuit and we did not consider flicker noise components. While the mechanisms that convert low-frequency $1/f$ noise into $1/f^3$ phase noise around the carrier are well understood and agreed upon, the derivation of explicit formulas for the computation of phase noise affected by flicker noise is still controversial. Different authors provide different expressions, and also different recipes for how to reduce its effects. The interested

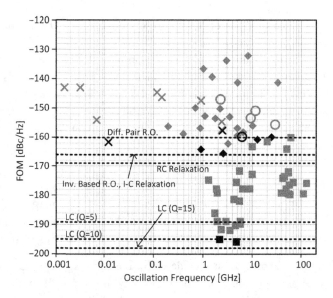

Figure 4.19 State of the art oscillator FOM vs. frequency: relaxation (crosses), differential pair based ring (circles), inverter-based ring (diamonds) and LC (squares). The best-in-class oscillators are indicated in black (see text for the references). The dashed lines are the theoretical FOM limits.

reader is referred to the papers cited in this chapter to have an overview of the different approaches to treat flicker noise conversion. We believe that the current status is not yet sufficiently consolidated to be inserted in this book at this time. On the other hand, in Section 9.4 we give a broad overview on the topic of $1/f$ noise in general, and some of the mathematical and conceptual difficulties connected to it. Flicker noise still remains to date an intriguing phenomenon, challenging the very mathematical tools on which our knowledge as engineers is built. The lack of a unified set of formulas to compute the effect flicker noise on oscillator phase noise might be, at least partly, due to it.

4.7   Ideal Frequency Divider

In this section we analyze how jitter of the output clock of an ideal frequency divider by N relates to the jitter of the input clock. In doing it, we will assume that the divider does not introduce jitter by itself, and that there is no delay between input and output clocks. The analysis of jitter and phase noise in differential CML-type frequency dividers has been thoroughly carried out in [71] and largely follows the same steps outlined to compute the jitter of a differential stage ring oscillator presented before. We will omit this analysis here and refer the reader to the original publication. Under the above-stated assumptions, the position of each output edge is perfectly aligned with that of the input edge which caused that specific transition in the output clock (see Figure 4.20, where $N = 4$ has been chosen). Since an output edge occurs every N input edges, the k-th

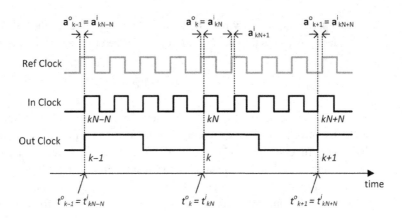

Figure 4.20 Ideal divider ($N = 4$).

output edge position t_k^o is aligned to the kN-th input edge: $t_k^o = t_{kN}^i$, where the indexes "i" and "o" denote input and output clocks respectively.

Following on from this simple observation, the position of the k-th output edge relative to the edge of an arbitrary reference clock, whether ideal or non-ideal, is identical to the position of the kN-th input edge relative to the same reference edge. This leads to the conclusion that the absolute (relative) jitter of the output clock is equal to the absolute (relative) jitter of the input clock, downsampled by the factor N:

$$\mathbf{a}_k^o := \mathbf{a}_{kN}^i \tag{4.102}$$
$$\mathbf{r}_k^o := \mathbf{r}_{kN}^i. \tag{4.103}$$

Since the absolute jitter is not affected by the divider but the clock period is multiplied by N, the output excess phase is equal to the input excess phase divided by N (refer to Equation 3.5). In terms of PSD, this leads to the conclusion that the output phase noise is N^2 times smaller than the input phase noise. Expressed in dB:

$$\mathcal{L}^o(f) := \mathcal{L}^i(f) - 20\log(N). \tag{4.104}$$

For instance, dividing by 2 lowers the phase noise profile by 6dB.

As far as the M-period jitter is concerned, following the same argument as above, the jitter between two output edges spaced M periods apart is equal to the jitter between the input edges which triggered them. The input edges, though, are spaced M times N input periods apart, so that the relation between input and output M-period jitter can be written as:

$$\mathbf{p}_k^o(M) = \mathbf{p}_{kN}^i(NM). \tag{4.105}$$

This relation can be also formally derived as follows:

$$\mathbf{p}_k^o(M) := t_{k+M}^o - t_k^o = t_{(k+M)N}^i - t_{kN}^i =: \mathbf{p}_{kN}^i(NM). \tag{4.106}$$

Therefore, as for absolute and relative jitter, the output N-period jitter is obtained by downsampling the input N-period jitter by a factor equal to the divider factor.

As a particular but very important case, the period jitter of the clock at the output of a divider by N is equal to the N-period jitter of the clock at the divider input:

$$\mathbf{p}_k^o = \mathbf{p}_{kN}^i(N). \tag{4.107}$$

As will be shown in the following chapters, in many typical applications, for instance a clock coming from a free-running oscillator or from a PLL, the M-period jitter is an increasing function of M. Combined with the equation above, this leads to the fact that the output period jitter is typically larger than the input period jitter.

The fact that an ideal frequency divider by N actually downsamples the input jitter by a factor of N has interesting consequences, analogous to those found in signal processing when decimating a discrete time signal.

The most prominent effect is connected to the case of an input jitter sequence exhibiting a repetitive, periodic behavior. This situation occurs often in practice, for instance, when the circuit producing the clock is disturbed by a periodic noise source, e.g., a ripple on the supply. In this case, the jitter will show the same periodicity as the periodic noise source.

Let us assume that the input clock is affected by absolute jitter periodic in five clock cycles, as shown in Figure 4.21 (top). If we divide this clock by 5, the output edges of the divided clock will be affected by an absolute jitter which is equal to the input jitter downsampled by a factor of 5. The bottom of Figure 4.21 shows the downsampling process. It can be seen that, independent of when the divider starts, the output jitter will be constant, and, thus, indistinguishable from a time offset. As a result the jitter is completely eliminated from the output clock.

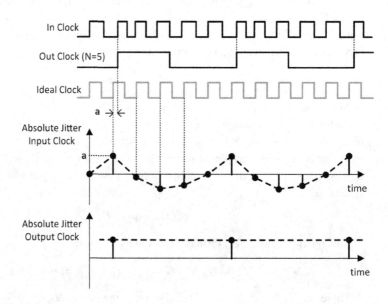

Figure 4.21 Jitter downsampling in an ideal frequency divider by $N = 5$.

Generally speaking, if a clock with absolute jitter containing a component periodic in P clock cycles is fed to a divider by kP, k being any integer, this jitter component is perfectly removed from the output clock.

If the divider factor is not equal to the periodicity of the jitter, the result is not immediate and is better analyzed in the frequency domain. If we denote the frequency of the input clock as f_0, and assume the jitter is sinusoidal with a periodicity of P clock cycles, then the spectrum of the jitter sequence (obtained via DFT, [72]) is given by two Dirac functions at $\pm f_0/P$, plus their repetitions around all integer multiples of f_0, and can be expressed as $nf_0 \pm f_0/P$, with n any integer. The spectrum of the jitter sequence downsampled by N is given by shifting the original spectrum by if_0/N, with $i = 0, 1, \ldots, N-1$ and summing up all the shifted versions. Therefore the spectrum of the downsampled jitter is composed of Dirac functions centered at:

$$nf_0 \pm \frac{f_0}{P} + i\frac{f_0}{N}, \quad n \in \mathbf{Z}, \quad i = 0, 1, \ldots, N-1. \tag{4.108}$$

Depending on the values of N and P, the original tone at f_0/P can alias into unwanted and unexpected jitter components. Figure 4.22 illustrates this behavior with three cases.

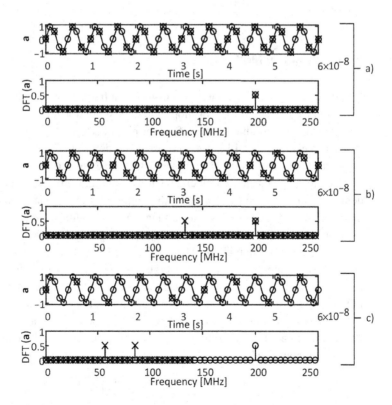

Figure 4.22 Jitter sequences and corresponding spectra for ideal dividers. For each case, the top graph shows the jitter sequence and the bottom graph the corresponding spectrum. The circles (crosses) indicate the values for the input (output) clock. The divider factor is: (a) $N = 2$, (b) $N = 3$, (c) $N = 7$.

The clock frequency f_0 is 1GHz and the absolute jitter is a sinusoid with period $P = 5$. The divider factor N is chosen to equal 2, 3, and 7, respectively. The figure shows the original jitter sequence and the downsampled one together with the correspondent spectra, obtained by applying a DFT to the jitter sequence. It can be seen that while the tone of the undivided jitter sequence is at $f_0/P = 200$MHz, the jitter of the divided sequence can show components at other frequency, as a result of the aliasing. For the first case ($N = 2$) the Nyquist criterion[7] is satisfied, so that the divided clock shows jitter at 200MHz. In the other two cases, Nyquist is not satisfied and tones are created at lower frequencies. For $N = 3$ the tone is at $-f_0/P + f_0/N = 133.33$MHz, while for $N = 7$ the tone is at $-f_0 + f_0/P + 6f_0/N = 57.149$MHz.

It is left as an exercise for the reader to prove that, depending on the value of the ratio N/P, the frequency of the jitter component at the output of the divider is given as:

$$f_{tone} = \begin{cases} \frac{f_0}{P} & \text{if } \frac{N}{P} <= \frac{1}{2} \\ -\frac{f_0}{P} + \lceil \frac{N}{P} \rceil \frac{f_0}{N} & \text{if frac}(\frac{N}{P}) \in [0.5, 1] \\ -f_0 + \frac{f_0}{P} + \lceil N - \frac{N}{P} \rceil \frac{f_0}{N} & \text{if } \frac{N}{P} > \frac{1}{2} \text{ and frac}(\frac{N}{P}) \in [0, 0.5] \end{cases} \quad (4.109)$$

where $\lceil x \rceil$ denotes the smallest integer greater or equal to x and frac(x) the fractional part of x.

4.8   Ideal Frequency Multiplier

While frequency division (at least at not-too-high frequencies) can be implemented with very simple circuits based on flip-flops, the same is not true for frequency multiplication. There is no easy or simple way of performing this operation, and more complicated circuits like PLLs or DLLs must be used, the descriptions of which are beyond the intended scope of this section. Here we will assume the existence of a machine capable of receiving a given jittered clock signal as input and providing at its output a clock defined such that each N-th edge of the output clock coincides with one edge of the input clock, and, between two such edges, $N - 1$ additional uniformly spaced edges are placed. In other terms, the edges of the multiplied clock are a linear interpolation of the edges of the input clock. The resulting clock output thus has a frequency which is N times the input frequency (see Figure 4.23). Even though this is an abstraction, it is essentially equivalent to the operation of a PLL, if bandwidth limitation issues, influencing the speed at which the output clock adapts to the input clock, are neglected.

Let us now consider how period, absolute and M-period jitter are transformed by such an ideal machine. Based on the mechanism described above, the relationship between output and input clock edge positions can be readily expressed as:

$$t^o_{kN+l} = t^i_k + \frac{l}{N}(t^i_{k+1} - t^i_k) \quad (4.110)$$

[7] The Nyquist criterion states that a signal with limited bandwidth B can be sampled without loss of information only if the sampling frequency is larger than $2B$. In the context discussed in this section, a clock affected by a jitter component periodic in P and then divided by N, the Nyquist criterion holds if $P > 2N$.

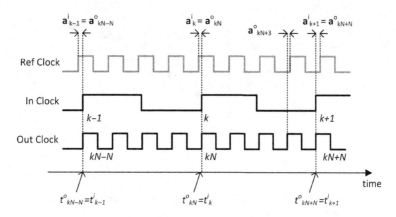

Figure 4.23 Time waveform for an ideal frequency multiplier by $N = 4$.

where t_k^o and t_k^i denote the k-th edges of the output and input clocks, respectively, and $l = 0, \ldots, N - 1$ is a running index. Based on this, it is easy to derive the expressions for jitter. The absolute jitter is indeed defined as:

$$\mathbf{a}_{kN+l}^o = t_{kN+l}^o - (kN + l)\frac{T}{N} \qquad (4.111)$$

where T is the nominal input clock period and therefore T/N is the nominal output clock period. By using Equation 4.110, the output absolute jitter can be expressed as a function of the absolute input jitter as:

$$\mathbf{a}_{kN+l}^o = (1 - \frac{l}{N})\mathbf{a}_k^i + \frac{l}{N}\mathbf{a}_{k+1}^i \qquad (4.112)$$

and the period jitter, obtained as the difference between two consecutive absolute jitter samples, simply becomes:

$$\mathbf{p}_{kN+l}^o = \frac{\mathbf{p}_k^i}{N} \qquad (4.113)$$

that is, the output period jitter is one N-th of that of the input.

In terms of phase noise, a similar argument as for the frequency divider applies, but with the opposite result. Since the absolute jitter is not affected by the multiplier but the clock period is *divided* by N, the output excess phase is equal to the input excess phase multiplied by N (refer to Equation 3.5). The output phase noise is thus N^2 times larger than the input phase noise. Expressed in dB:

$$\mathcal{L}^o(f) := \mathcal{L}^i(f) + 20\log(N). \qquad (4.114)$$

For instance, an ideal multiplier by 2 increases the phase noise profile by 6dB.

As far as the M-period jitter is concerned, we can argue that the jitter over M consecutive periods can be decomposed into the sum of the jitter over the first $N\lfloor M/N \rfloor$ periods and the jitter over the remaining $M - N\lfloor M/N \rfloor$ periods. The operator $\lfloor \cdot \rfloor$ indicates the largest integer smaller than or equal to the operand. Since $N\lfloor M/N \rfloor$ periods of the output clock coincide with $\lfloor M/N \rfloor$ periods of the input clock, the first contribution

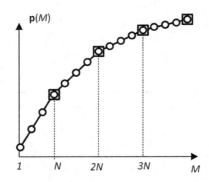

Figure 4.24 M-period jitter for an ideal frequency multiplier by N for the input (squares) and for the output (circles) clock.

is equal to the $\lfloor M/N \rfloor$-period jitter of the input clock. By virtue of Equation 4.113, the second contribution is equal to $(M - N \lfloor M/N \rfloor)$ times the input period jitter divided by N. Combining these two contributions, the equation for the output M-period jitter can be written as (note that in the following formula the brackets in the first term on the right-hand side are the argument of the N-period jitter; it is not a multiplication):

$$\mathbf{p}^o(M) = \mathbf{p}^i \left(\left\lfloor \frac{M}{N} \right\rfloor \right) + \left(\frac{M}{N} - \left\lfloor \frac{M}{N} \right\rfloor \right) \cdot \mathbf{p}^i(1). \qquad (4.115)$$

Seen from a different perspective, every N-th output edge coincides with one input edge, and therefore the kN-period jitter on the output clock is equal to the N-period jitter of the input one, with k an integer between 1 and $N - 1$. Since the edges in between are the result of a linear interpolation, the other samples of the output M-period jitter will be the linear interpolation between two adjacent samples of that of the input (see Figure 4.24).

5 Effects of Jitter in Synchronous Digital Circuits

In this chapter we analyze the effect of jitter on synchronous digital systems. We will first introduce the register (also known as flip-flop) as the central element of the design, its associated timing requirements, and the standard configuration for edge-triggered digital design. After analyzing the effect of jitter on the standard configuration, the reader will be introduced to the case of divided clock systems, enabled systems and multicycle systems. Finally, a section is dedicated to latch-based digital design and the effect of jitter on it.

5.1 Edge-Triggered Synchronous Design

In modern integrated systems, the vast majority of the digital functions are implemented in edge-triggered synchronous design. The central element of these systems is the *register*, a circuit which transfers to its output the digital value of its input, on the occurrence of the edge of a clock signal. Shown in Figure 5.1 is the symbol of a register and the corresponding timing waveforms. In an ideal register, when the rising edge of the clock signal CK occurs, the output Y assumes immediately the value of the input X at that particular instant. In practice, a register behavior differs from the ideal in at least three aspects. First, in order to be able to capture the input data correctly, the input data has to be stable some time before the rising edge of CK. This time is called *setup time* and is indicated by τ_{su}. Second, the input data is not allowed to change its value for some time after the rising edge of CK. This time is called the *hold time* and is indicated by τ_{ho}. Third, even if setup and hold time constraints are satisfied, there is a propagation delay from input to output, so that the new data appears at the output Y some time after the rising edge of CK. This time is typically called the *clock-to-Q*, and is indicated by τ_{cq}.

Figure 5.2 illustrates the typical configuration encountered in synchronous edge-triggered designs. A first register, or bank of registers, samples the incoming data X1 with the clock CK and produces the output Y1. This output is processed by a combinatorial logic network, and its result X2 is fed to the input of a second register, or bank of registers. The second register samples the processed data X2 with the same clock CK, and produces the output Y2.

Let us analyze this configuration from the point of view of the setup time constraint. Once the data X1 is sampled on one edge of CK, the output Y1 changes after τ_{cq}. The new Y1 data goes through the combinatorial logic and X2 changes after a time

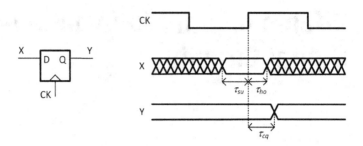

Figure 5.1 Illustration of basic register timing.

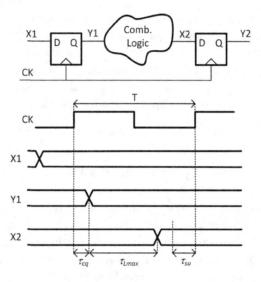

Figure 5.2 Timing diagram of clocking in a edge-triggered synchronous design.

which depends on the delay introduced by the combinatorial logic. To avoid errors in the second register, X2 must be stable at least τ_{su} before the next CK edge. The worst case scenario occurs when the delay of the combinatorial logic is maximum, τ_{Lmax}. Collecting all terms, the condition on timing for proper operation of the system can be written as:

$$T > \tau_{cq} + \tau_{Lmax} + \tau_{su} \tag{5.1}$$

imposing a bound on the minimum duration (maximum frequency) of the clock period T.

Regarding the hold time constraint, the data at the input of the second register is not allowed to change for at least τ_{ho} after the rising edge of CK. If we assume for now that the rising edges on the first and second register are perfectly synchronous, the data X2 will change $\tau_{cq} + \tau_{Lmin}$ after the CK edge, where for the worst case scenario we assumed the minimum combinatorial delay τ_{Lmin}. Translating this into equations, the condition for the hold time to be satisfied is:

$$\tau_{cq} + \tau_{Lmin} > \tau_{ho}. \tag{5.2}$$

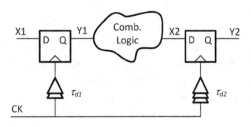

Figure 5.3 Illustration of skew in synchronous design.

It can be seen that the hold time does not impose any constraints on the frequency of the clock, only on the minimum delay of the combinatorial logic. This is because the hold time refers to the *same* edge of CK, while the setup time involves two consecutive edges of CK.

There are two main factors which affect the ideal scenario depicted above: clock skew and clock jitter. Clock skew occurs when the clock distribution to the registers is not perfectly balanced and the CK edge arrives at slightly different instants at different registers. This phenomenon is typical in big digital cores, due to extremely large and complex clock distribution networks, and is one of the most important performance degradation factors. Figure 5.3 illustrates the issue in a simplified manner. The clock CK is distributed to the registers using a clock tree made of buffers. Even though modern tools try to balance the delay of CK from the common source to each single register, there are always some residual delay differences. In the case in figure, if τ_{d1} does not match τ_{d2}, the two registers experience a clock skew. Indicating with τ_{sk} the skew delay $\tau_{d2} - \tau_{d1}$, it is easy to prove that the timing constraints for setup and hold time translate into:

$$T > \tau_{cq} + \tau_{Lmax} + \tau_{su} - \tau_{sk} \tag{5.3}$$

$$\tau_{Lmin} > \tau_{ho} - \tau_{cq} + \tau_{sk}. \tag{5.4}$$

In the following, we analyze the effects of jitter on the timing constraints of the system shown in Figure 5.3. We assume that the clock distribution network does not introduce any jitter in addition to the one on clock CK. Starting from the case of the setup time, the relevant aspect is the position in time of the clock edge sampling the second register, with respect to the previous clock edge sampling the first register. These two clock edges are consecutive so that the relevant jitter is the period jitter. It can be understood that if the period jitter is too high (in absolute value), the clock period might be much shorter than expected, leading to a violation of the setup time constraint. In a more formal way, we can replace the period T with the sum of the nominal period T_{nom} and the period jitter **p** in Equation 5.3, so that the constraint on the period jitter becomes:

$$\mathbf{p} > \tau_{cq} + \tau_{Lmax} + \tau_{su} - \tau_{sk} - T_{nom}. \tag{5.5}$$

In a properly working design, the right-hand side of this equation is a negative number, so that the equation above impose a minimum limit to the period jitter. Recalling the definition of negative peak period jitter in Equation 2.34, the constraint on jitter can be written as:

$$\rho_{\mathbf{p}}^{-} < T_{nom} + \tau_{sk} - (\tau_{cq} + \tau_{Lmax} + \tau_{su}). \qquad (5.6)$$

For the hold time case, since the clock edge involved in the timing constraint is the very same for the first and the second register, and we assumed that no additional jitter is introduced by the clock distribution network, jitter on the clock will have no net effect on the timing margin. In other words, the hold time does not impose any constraint on the clock jitter.

When **p** has a random jitter component, its distribution is Gaussian and Equation 5.6 cannot be satisfied. If the random jitter RMS value is small compared to the nominal timing margin, the probability of a setup time violation will be small, but can never be zero and the register experiences what is called a *metastable state*. In this state, the internal circuitry of the register does not have enough gain to sample correctly the input data and to transfer it to the output in a predictable time. It could happen that either the wrong data is produced at the output, or, even if the correct data is sampled, the output reaches a usable voltage level only after a very long time, introducing errors in the following logic. The analysis of the effect of metastability on the performance of digital systems involves statistical and probabilistic considerations and ultimately leads to the concept of mean time between failures (MTBF). The MTBF is an indication of how often errors in digital systems occur because of metastability and, for a single register with asynchronous relation between data and clock, can be expressed as:

$$\text{MTBF} = \frac{e^{(t_r/\tau_{BW})}}{f_D f_{CK} T_w}. \qquad (5.7)$$

In this equation, f_D is the frequency of the input data, f_{CK} is the frequency of the clock, T_w and τ_{BW} are two parameters determined by the specific implementation of the register (the first is the metastability window, the second is the inverse of the gain-bandwidth product of the register), and finally t_r is the time (called the *resolution time*) allowed by the system to bring the output to a valid level.

The rate of tolerable violations in a specific system depends, among other factors, on the number of registers, on the availability of recovery mechanisms, on how sensitive each subsystem is to synchronization error. A discussion of the effects of metastability on systems is beyond the scope of this book, and the interested reader can refer to [73], [74], [75], and [76] for a deeper insight and further references.

5.2   Gated Clock, Divided Clock, Enabled Systems

In digital designs, not all registers are always clocked at the same frequency. Figure 5.4 illustrates in a simplified manner three of the most common cases. In one particular implementation, some parts of the design may be allowed to run at a lower speed, with registers clocked by a divided version of the main clock, as shown in Figure 5.4 A. In other implementations, data are clocked in a discontinuous and almost "on demand" fashion, rather than in a continuous mode. In such cases, either the clock is gated, allowing transitions to reach the register only when data need to be sampled, as shown in

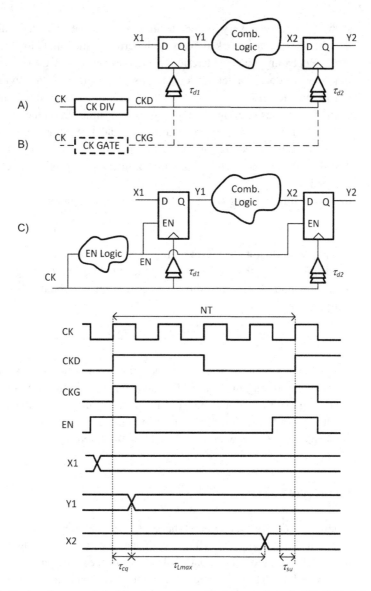

Figure 5.4 Timing diagram of clocking in a system with: (A) gated clock, (B) divided clock, and (C) enable signal.

Figure 5.4 B, or a signal enables the registers only in specific moments (Figure 5.4 C). Note that the cases A and B are to be considered mutually exclusive. They are depicted in the same figure just for illustration purposes. It is important to note that, independent of which scheme has been used, the result is that the registers are clocked every N cycles of the input clock, rather than on consecutive edges.

Let's now analyze the effects of jitter on clock CK on the timing constraints of the system shown in Figure 5.4, assuming that the divider does not introduce additional skew.[1] This case is very similar to the previous one, with the only difference being that the registers are clocked every N-th period of the input clock CK, instead of every clock. Repeating the consideration of the previous case we can conclude that the hold time closure does not set any constraints on the jitter of the divided or gated clocks (CKD and CKG respectively), or on the clock CK. As far as the setup time is concerned, from the previous section it follows that the constraint is on the peak period jitter of the divided or gated clocks (for divided or gated clock systems), or on the displacement of one edge of the clock CK with respect to the N-th previous edge (for enabled systems). Translating these requirements in terms of jitter on the original clock CK, they all imply a bound on the peak N-period jitter of the clock CK. Recalling Equation 5.6, we can then write the following for the systems under investigation:

$$\rho_{\mathbf{p}(N)}^{-} < NT_{nom} + \tau_{sk} - (\tau_{cq} + \tau_{Lmax} + \tau_{su}). \tag{5.8}$$

It is important to note that N-period jitter is typically larger than period jitter. As shown in Section 3.2.2, in the case of a clock generated by a free-running oscillator or by a PLL with sufficiently low bandwidth, the N-period jitter is about $\sqrt{N}$ larger than the period jitter, so that the equation above can be rewritten as:

$$\rho_{\mathbf{p}}^{-} < \frac{NT_{nom} + \tau_{sk} - (\tau_{cq} + \tau_{Lmax} + \tau_{su})}{\sqrt{N}}. \tag{5.9}$$

If the digital system is designed to pack as much logic as possible between any two registers, the timing margin at the numerator of this equation is of the same order as the timing margin available between two registers clocked by the undivided clock. This results in $\sqrt{N}$ more stringent conditions on the period jitter of the undivided clock, compared to a system where no divided clock is used.

5.3   Multicycle Paths

In complex designs, some combinatorial logic may take longer than one clock period to produce its result. If the function of the system can tolerate a delay of more than one clock cycles on these paths, there is no need to change the architecture of the design, and this path can be treated as a multicycle path. A multicycle path over N clock cycles is a combinatorial path which is expected to produce its results between the next $(N - 1)$-th and N-th edge of the clock. The diagram of Figure 5.5 illustrates an example of multicycle path for $N = 3$. It is important to note that, unlike in divided or gated clock systems, in multicycle paths the second register is always enabled and clocked at every edge of the clock CK. We will see that this makes a difference in constraining for jitter.

We want now to analyze the effects of clock jitter on the timing constraints of the system shown in Figure 5.5. Starting from the analysis of setup time, the data at the output of the combinatorial logic must be stable at the input X2 of the second register

---

[1] In practical cases the divider will introduce a static, deterministic skew between the divided and the non-divided clocks. This skew has to be taken into account in the timing budget of the timing arcs preceding and following those clocked by the divided clock.

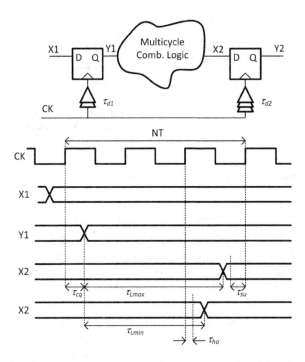

Figure 5.5 Timing diagram of clocking in a synchronous system with multicycle paths.

τ_{su} before the N-th edge of clock CK. From this point of view there is no difference between this system and a divided or gated clock system, and Equation 5.8, constraining the peak N-period jitter, applies.

An additional bound on jitter is in this case determined by the hold time requirements. Indeed, the data at the output of the combinatorial logic must arrive at the input X2 τ_{ho} after the $(N - 1)$-th edge of clock CK. If this is not guaranteed, the data might appear at the output Y2 one clock cycle too early. Following the steps outlined in the previous cases it is left as an easy exercise for the reader to prove that this condition impose a bound on the maximum *positive* peak $(N - 1)$-period jitter:

$$\rho_{p(N-1)}^+ < \tau_{cq} + \tau_{Lmin} - \tau_{ho} - [(N - 1)T_{nom} + \tau_{sk}] \qquad (5.10)$$

where τ_{Lmin} is the *minimum* propagation delay of the combinatorial logic.

5.4   Latch-Based Synchronous Design

A latch is a basic digital circuit often used in full custom digital systems instead of a register. While a register samples the input data on the *edge* of the clock, a latch is sensitive to the *level* of the clock. When the clock is high, the output is equal to the input. If during the high clock phase the input changes, the output changes accordingly. In this phase the latch is said to be *transparent*. When the clock goes low, the output data retains the latest value assumed. In this phase, if the input changes, the output doesn't,

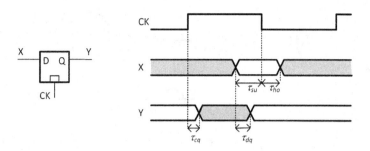

Figure 5.6 Illustration of basic latch timing.

and the latch acts as an isolation element between input and output. The use of latches in a digital design can potentially give rise to dangerous race conditions, difficult to spot in a complex system, and is therefore not recommended in an automated RTL digital flow. Nevertheless, for full custom digital logic, the use of latches allows substantial gain in terms of maximum speed at which the system can work, and is therefore used for high-speed designs.

Figure 5.6 illustrates the symbol for a latch (note the difference between the symbols of the register and of the latch in correspondence of the clock input), together with its timing diagram. There are four main timing parameters associated with a latch. The setup and hold times (τ_{su} and τ_{ho}) are associated with the *falling* edge of the clock CK, rather than with the *rising* edge, as was the case with a register. In analogy to the register, the time it takes for the output to change when the latch goes into transparent mode is called *clock-to-Q*, τ_{cq}. When the latch is already in transparent mode and the input changes, the output changes after a propagation delay called *data-to-Q*, τ_{dq}.

In Figure 5.7 the basic structure of a latch-based synchronous logic is sketched. A first latch, transparent on the high level of clock CK, is followed by a first combinatorial logic block. This logic block is followed by a second latch which is clocked by the inverted clock CK, symbolized by a circle on the clock input pin. Therefore this latch is transparent during the low level of CK. A second combinatorial logic block follows, and a third latch, transparent on the high level of the clock as the first one, concludes the stage. Note how the use of latches in logic paths is nothing more than inserting combinatorial logic between the master and the slave stages of a master–slave register.

We will now analyze the effects of clock jitter on the timing constraints of the system shown in Figure 5.7. Starting from the setup time, let us assume that the first latch has no timing issues, so that data X1 is stable at least τ_{su} before the falling edge of CK. If this condition is satisfied, the output Y1 changes τ_{dq} after X1. The first combinatorial block processes this data and produces its result at the input X2 after the worst case propagation delay τ_{Lmax1}. In order to be properly latched by the second latch, the data must be stable at X2 at least τ_{su} before the next rising edge of CK (remember that the second latch has an inverted clock input). Collecting all this information and with the help of Figure 5.7, the following equation can be written for the setup constraining of the second latch:

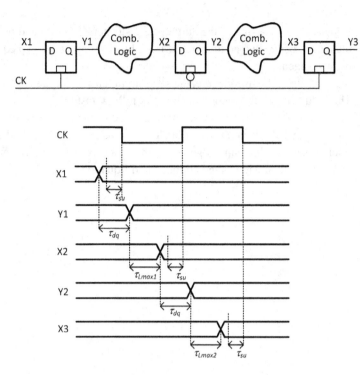

Figure 5.7 Timing diagram of clocking in a latch-based synchronous design.

$$0 - \tau_{su} + \tau_{dq} + \tau_{Lmax1} < T_{low} - \tau_{su} \tag{5.11}$$

where time zero has been chosen to correspond to the first falling edge of CK, and T_{low} indicates the duration of the low phase of CK. Note how in this case the setup time τ_{su} appears on both sides of the equation with the same sign, and thus can be eliminated from the timing budget. This is a significant difference with respect to edge-triggered systems. A second important observation is that the timing is sensitive to the duration of the low phase of clock CK and thus to the duty-cycle of the clock CK. Indicating with δ the duty-cycle of CK, with T_{nom} the nominal period, and with $\mathbf{p}(1/2)$ the $(1/2)$-period jitter, the duration of the low phase can be expresses as $(1 - \delta)T_{nom} + \mathbf{p}(1/2)$. Replacing this expression in the previous equation, the constraints on the jitter of one half-period can be expressed as:

$$\rho_{\mathbf{p}(1/2)}^- < (1 - \delta)T_{nom} - (\tau_{dq} + \tau_{Lmax1}). \tag{5.12}$$

The analysis of the path going from the second register to the third register is identical to the one above, with the only difference being that in this case the CK is negated. Therefore, instead of the low phase, the high phase of CK must appear in the equation, which can be thus written as:

$$\rho_{\mathbf{p}(1/2)}^- < \delta T_{nom} - (\tau_{dq} + \tau_{Lmax2}). \tag{5.13}$$

Depending on the sign of the duty-cycle distortion ($\delta - 0.5$) and on difference between the delays of the two combinatorial logic blocks, one of the two equations above is the most stringent.

As for the hold time closure, the same arguments as for the edge-triggered logic hold. The hold time on the second latch refers to how fast the data at its input changes after the falling edge of the clock CK. The falling edge on the second latch is generated by the same rising CK edge on the first latch, which releases the data to the second latch. Any jitter present on the rising edge of CK thus cancels out from the timing budget, exactly as for the case of edge-triggered logic, and Equation 5.4 holds.

6 Effects of Jitter on Data Converters

By virtue of being clocked, all data converters are susceptible to clock jitter. This chapter provides several case studies that include current DACs, Nyquist data converters, time-interleaved ADCs, and oversampling data converters. In each case, we first provide a brief background of the area to clearly define the problem we are trying to solve, then demonstrate how the concepts and techniques introduced in previous chapters could be used to analyze the effect of jitter and phase noise on circuit performance.

6.1 Effects of Jitter on Current DACs

6.1.1 Background

A binary current DAC (IDAC), as shown symbolically in Figure 6.1, takes a discrete-time input signal $x[k]$ and produces a continuous-time current signal $y(t)$ as its output. In addition to the input signal, the IDAC also receives a clock signal with frequency f_s, ideally with no jitter, and a reference current I_R. Accordingly, an ideal IDAC converts binary data, -1 or $+1$, to a current pulse with T_s duration and $-I_R$ or $+I_R$ amplitude, respectively. The relationship between the input and output of a binary IDAC can be written as:

$$y(t) = I_R \sum_k x[k](u(t - kT_s) - u(t - (k+1)T_s)) \qquad (6.1)$$

where T_s represents the ideal clock period and $u(t)$ represents the unit step function. A simplified circuit diagram for a binary IDAC is shown in Figure 6.2. Since the D input of the latch is assumed to be constant with sufficient margin beyond the setup and hold times of the latch, the sampling process is not influenced by the clock jitter. For this reason, the continuous-time input $x(t)$ could be represented by the discrete-time sequence $x[k] \in \{-1, +1\}$. A $+1$ results in a positive current $+I_R$ at the output (for the duration of one clock cycle) and a -1 results in a negative current $-I_R$.

We now consider the relationship between the clock jitter and the noise it will add to the output of the DAC.

6.1.2 Non-Return-to-Zero (NRZ) IDAC

Consider a non-ideal binary IDAC, as shown in Figure 6.2, with a jittery clock CK. We are interested in deriving an expression for the noise power at the output of the IDAC due to the clock jitter.

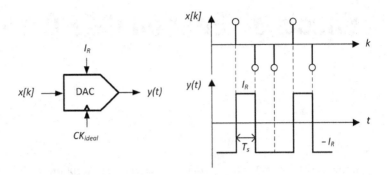

Figure 6.1 Block diagram of an IDAC along with a sketch of its input and output signals.

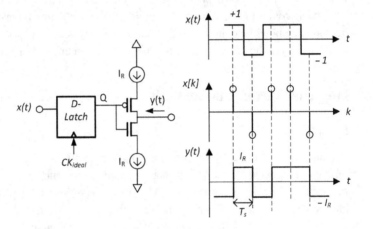

Figure 6.2 A simplified circuit diagram of a unit element along with its input and output waveforms.

As shown in Figure 6.3, the output $y_{id}(t)$ of an ideal IDAC consists of a waveform that is either $+I_R$ or $-I_R$. Therefore, the output signal power is I_R^2. The output $y(t)$ of a non-ideal IDAC deviates from that of an ideal IDAC only at output transition times. If we denote by $\mathbf{a}_k$ the absolute jitter of the jittery clock, then the difference between the two outputs, which we define as the error, can be written as:

$$e(t) = y(t) - y_{id}(t) = \sum_k e_k(t) \tag{6.2}$$

where

$$e_k(t) = -2I_R \frac{x[k] - x[k-1]}{2} \, \text{sgn}(\mathbf{a}_k) \, (u(t - kT_s) - u(t - kT_s - \mathbf{a}_k)) \tag{6.3}$$

where sgn($\cdot$) represents the *sign* function, which produces -1, $+1$, or 0, depending on whether its argument is negative, positive, or zero, respectively. If we further define $\Delta_k = (x[k] - x[k-1])/2$, then $\Delta_k \in \{+1, -1, 0\}$ represents a positive transition, a negative transition, or no transition in the input data sequence. We can then write:

$$e_k(t) = -2I_R \, \Delta_k \, \text{sgn}(\mathbf{a}_k) \, (u(t - kT_s) - u(t - kT_s - \mathbf{a}_k)). \tag{6.4}$$

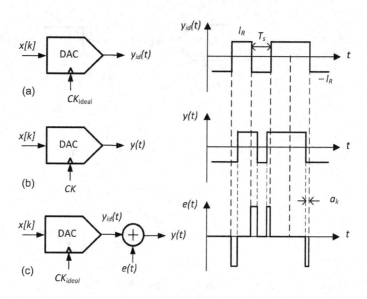

Figure 6.3 (a) An IDAC with an ideal clock along with its output waveform, (b) an IDAC with jittery clock along with its output waveform, (c) a model of a DAC with jittery clock.

To find the power of the error signal, σ_e^2, we first find the average power of $e_k(t)$ over one period as a function of k and then find the expected value of this average power.

$$e_k^2(t) = 4\, I_R^2 \Delta_k^2 (u(t - kT_s) - u(t - kT_s - \mathbf{a}_k)) \tag{6.5}$$

$$e_k^2 = \frac{1}{T_s} \int_{T_s} e_k^2(t)\,dt = 4\, I_R^2 \Delta_k^2 \frac{|\mathbf{a}_k|}{T_s} \tag{6.6}$$

$$\sigma_e^2 = E[e_k^2] = 4\, I_R^2\, \sigma_\Delta^2 \frac{E[|\mathbf{a}_k|]}{T_s}. \tag{6.7}$$

For a random binary sequence $x[k] \in \{+1, -1\}$, $\sigma_\Delta^2 = 0.5$. Also, for a Gaussian random jitter $\mathbf{a}_k$ with a mean of zero and a standard deviation of σ_a, it can be shown easily that $E[|\mathbf{a}_k|] = \sqrt{\frac{2}{\pi}}\sigma_\mathbf{a}$. As a result, Equation 6.7, can be simplified to:

$$\sigma_e^2 = \sqrt{\frac{8}{\pi}}\, I_R^2\, \frac{\sigma_\mathbf{a}}{T_s}. \tag{6.8}$$

It is common to write the noise power in dBFS, which is defined as the noise power in dB relative to the power of the full-scale signal (I_R^2 in this case). Therefore, we can write:

$$\sigma_e^2 = 10\log\left(\sqrt{\frac{8}{\pi}}\,\frac{\sigma_\mathbf{a}}{T_s}\right) \quad \text{[dBFS]}. \tag{6.9}$$

It may appear odd that the error variance at the output of the DAC is proportional to the standard deviation of the jitter, and not to the variance of the jitter. The error variance as shown in this equation represents the total noise power due to jitter including both the in-band and the out-of-band noise. As we will see later in this section, the in-band portion of this noise power is proportional to the jitter variance.

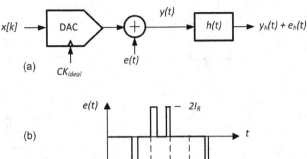

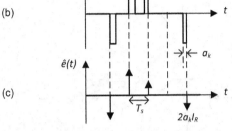

Figure 6.4 (a) A DAC followed by a low-pass filter, (b) a sketch of an error waveform, and (c) the error waveform approximated by delta functions.

6.1.3   NRZ IDAC Followed by a Linear Filter

Assume the DAC is followed by a linear filter as shown in Figure 6.4(a). We are interested in deriving an expression for the noise at the output of the linear filter due to clock jitter.

If we denote by $h(t)$ and $s(t)$ respectively the impulse response and the step response of the linear filter, we can write the following expression for the error at the output of the linear filter, $e_h(t)$:

$$e_h(t) = -2I_R \sum_k \Delta_k \, \text{sgn}(\mathbf{a}_k) \, (s(t - kT_s) - s(t - kT_s - \mathbf{a}_k)). \qquad (6.10)$$

If we assume $|\mathbf{a}_k| \ll T_s$, and note that $h(t) = ds(t)/dt$, we can use the Taylor expansion of $s(t)$ in the vicinity of $t = kT_s$, and write:

$$e_h(t) = -2I_R \sum_k \Delta_k \mathbf{a}_k h(t - kT_s)$$

$$= h(t) * -2I_R \sum_k \Delta_k \mathbf{a}_k \delta(t - kT_s). \qquad (6.11)$$

Note that this expression is valid only when the DAC is followed by a linear filter whose impulse response $h(t)$ is continuous at $t = 0$. If we now define $\hat{e}(t)$ as:

$$\hat{e}(t) := -2I_R \sum_k \Delta_k a_k \delta(t - kT_s) \qquad (6.12)$$

then we can write:

$$e_h(t) = h(t) * \hat{e}(t). \qquad (6.13)$$

This equation implies that, as far as $e_h(t)$ is concerned, we can substitute $e(t)$, as shown in Figure 6.4(b), with $\hat{e}(t)$, as shown in Figure 6.4(c). This substitution, as we will see shortly, will simplify the derivation of the noise power at the output of the filter. $\hat{e}(t)$ is a cyclostationary process since its statistical properties repeat every T_s. The autocorrelation function of $\hat{e}(t)$ can be written as follows:

$$R_{\hat{e}}(t_1, t_2) = 4I_R^2 \sum_j \sum_k E[a_j \Delta_j a_k \Delta_k] \delta(t_1 - jT_s) \delta(t_2 - kT_s)$$

$$= 4I_R^2 \sum_n R_a[n] R_\Delta[n] \sum_k \delta(t_1 - (n+k)T_s) \delta(t_2 - kT_s) \qquad (6.14)$$

where we have assumed that a_k and Δ_k are independent, and defined $n = j - k$. Rewriting this equation for $t_1 = t + \tau$ and $t_2 = t$, will result in:

$$R_{\hat{e}}(t + \tau, t) = 4I_R^2 \sum_n R_a[n] R_\Delta[n] \sum_k \delta(t + \tau - (n+k)T_s) \delta(t - kT_s) \qquad (6.15)$$

This equation clearly shows that the autocorrelation is a function of both t and τ, as is the case for any cyclostationary process. However, if we define $\bar{e}(t) = \hat{e}(t - \phi)$, where ϕ is a uniformly distributed random phase in $(0, T_s)$, then it can be shown [1] that:

$$R_{\bar{e}}(\tau) = \frac{1}{T_s} \int_0^{T_s} R_{\hat{e}}(t + \tau, t) dt \qquad (6.16)$$

where autocorrelation is a function of τ only. Using $\bar{e}(t)$ instead of $\hat{e}(t)$ to represent the error is justified because the phase difference between the input and the clock is not known at time zero. Substituting Equation 6.15 in Equation 6.16 results in:

$$R_{\bar{e}}(\tau) = \frac{4I_R^2}{T_s} \sum_n R_a[n] R_\Delta[n] \delta(\tau - nT_s). \qquad (6.17)$$

Taking the Fourier transform of both sides, we now find the power spectral density (PSD) of the error signal at the DAC output:

$$S_{\bar{e}}(f) = \frac{4I_R^2}{T_s} \sum_n (R_a[n] R_\Delta[n]) e^{-j\omega n T_s}. \qquad (6.18)$$

If we further assume that a_k and Δ_k are white, then their autocorrelation functions will be zero except when $n = 0$. Therefore, we can write:

$$S_{\bar{e}}(f) = \frac{4I_R^2}{T_s} \sigma_a^2 \sigma_\Delta^2. \qquad (6.19)$$

Note that this equation approximates the spectral density of the error under the assumption that $a_k \ll T_s$, where error pulses could be approximated by delta functions. For a jitter that does not satisfy this condition, the above equation is only valid at lower frequencies.

To find the total error power at the output of the filter, we must multiply this spectral density by $|H(j\omega)|^2$ and integrate over all frequencies. If we assume the linear filter is an ideal low-pass filter with a flat frequency response for $-f_0 < f < f_0$, and zero outside its passband, we can write:

$$\sigma_{\bar{e}_h}^2 = 2f_0 S_{\bar{e}}(f) = 4I_R^2 \sigma_\Delta^2 \left(\frac{\sigma_a}{T_s} \right)^2 \frac{1}{OSR} \qquad (6.20)$$

where OSR is the oversampling ratio, and is defined as $OSR = f_s/2f_0$. This expression provides the in-band noise power due to jitter and is indeed proportional to the jitter variance.

6.1.4   NRZ IDAC Followed by an Integrating Capacitor

Assume the IDAC is followed by an integrating capacitor as shown in Figure 6.5. We are interested in deriving an expression for the power of the current noise that is delivered to the capacitor as result of DAC clock jitter.

The charge error provided to the capacitor in the k-th cycle can be written as:

$$Q_e(kT_s) = 2I_R\Delta_k\mathbf{a}_k. \tag{6.21}$$

Since this charge is delivered every T_s seconds, the average error current per cycle can be written as:

$$\bar{e}(kT_s) = 2I_R\Delta_k\left(\frac{\mathbf{a}_k}{T_s}\right). \tag{6.22}$$

Again, assuming Δ_k and $\mathbf{a}_k$ are uncorrelated, we can find the following for the error power:

$$\sigma_{\bar{e}}^2 = 4I_R^2\sigma_\Delta^2\left(\frac{\sigma_\mathbf{a}}{T_s}\right)^2. \tag{6.23}$$

To express this power in dBFS, we will divide it by I_R^2. As before, we further assume $\sigma_\Delta^2 = 0.5$. Therefore, we can write:

$$\sigma_{\bar{e}}^2 = 10\log\left(2\left(\frac{\sigma_\mathbf{a}}{T_s}\right)^2\right) \quad \text{[dBFS]}. \tag{6.24}$$

This result is consistent with Equation 6.20 if $OSR = 1$.

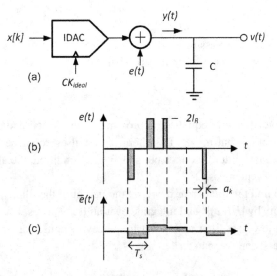

Figure 6.5 (a) An IDAC followed by an integrating capacitor, (b) a sketch of an error waveform, and (c) the per-cycle average error waveform.

6.1.5   Return-to-Zero IDAC

A Return-to-Zero (RZ) binary IDAC, as shown in Figure 6.6, receives as its input a binary sequence and produces as its output a CT current according to the following equation:

$$y(t) = I_R \sum_k x[k](u(t - kT_s - \alpha - \mathbf{a}_k) - u(t - (k+1)T_s - \beta - \mathbf{b}_k)) \qquad (6.25)$$

where α and β (both less than T_s) represent the nominal edge locations of the DAC pulse and $\mathbf{a}_k$ and $\mathbf{b}_k$ represent two independent random processes corresponding to the absolute jitter of the two edges. We are interested in finding the power of the error produced by jitter.

As shown in Figure 6.6, an ideal output, $y_{id}(t)$ corresponds to the case where $\mathbf{a}_k = \mathbf{b}_k = 0$. As a result, the error signal, defined as $e(t) = y(t) - y_id(t) = \sum_k e_k(t)$, where $e_k(t)$ can be written as follows:

$$e_k(t) = -I_R[(u(t - kT_s - \alpha) - u(t - kT_s - \alpha - \mathbf{a}_k))$$
$$+ (u(t - kT_s - \beta) - u(t - kT_s - \beta - \mathbf{b}_k)). \qquad (6.26)$$

This error waveform is different from that of the NRZ DAC in two ways: first, there are always two transitions per DAC cycle, irrespective of whether the input data has a transition. This eliminates the need to include Δ_k as defined for the case of NRZ DAC. Second, the magnitude of the error waveform is now I_R instead of $2I_R$. The procedure to find the power of error in case of RZ DAC is similar to the NRZ case and can be derived easily by the reader. Here, we suffice to include the final results corresponding to each

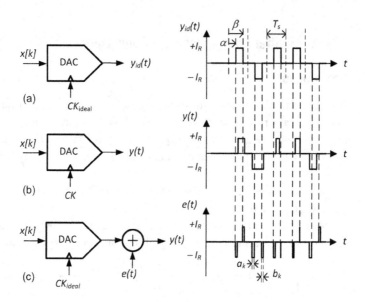

Figure 6.6 (a) An RZ IDAC with an ideal clock along with its output waveform, (b) an RZ IDAC with a jittery clock along with its output waveform, (c) a model of an RZ DAC with a jittery clock.

of the three cases we discussed earlier. In all three cases, we assume $\alpha = 0.25T_s$ and $\beta = 0.75T_s$, corresponding to a DAC pulse that is half of the clock period.

Corresponding to Equation 6.9, the error power in dBFS at the output of an RZ DAC is given by:

$$\sigma_e^2 = 10\log\left(\sqrt{\frac{2}{\pi}} \frac{\sigma_\mathbf{a} + \sigma_\mathbf{b}}{T_s}\right) \quad [\text{dBFS}]. \tag{6.27}$$

Corresponding to Equation 6.20, the total power of error (due to jitter) at the output of an RZ IDAC is given by:

$$\sigma_{\bar{e}_h}^2 = 2f_0 S_{\bar{e}}(f) = I_R^2 \left(\frac{\sigma_\mathbf{a}^2 + \sigma_\mathbf{b}^2}{T_s^2}\right) \frac{1}{OSR}. \tag{6.28}$$

And, corresponding to Equation 6.24, the error power of an RZ DAC followed by an integrator and an ideal sampler can be written in dBFS as:

$$\sigma_{\bar{e}}^2 = 10\log\left(\frac{\sigma_\mathbf{a}^2 + \sigma_\mathbf{b}^2}{T_s^2}\right) \quad [\text{dBFS}]. \tag{6.29}$$

Finally, for a fair comparison between the noise power of the RZ and the NRZ DACs, we must take into account the size of their full-scale currents. While we have denoted with I_R the full-scale current for both, the full-scale current of the RZ DAC should be considered larger than that of the NRZ DAC so as to produce the same full-scale voltage at the DAC output, assuming the same load capacitance. For the case of a RZ DAC with $\alpha = 0.25T_s$ and $\beta = 0.75T_s$, the RZ full-scale current amplitude should be twice as large as that of the NRZ DAC.

6.2   Effects of Jitter on Nyquist Data Converters

6.2.1   Background

An ADC samples a continuous-time analog signal, $x(t)$, at constant time intervals, T_s, to produce a corresponding discrete-time digital output signal, $y[k]$. Due to limited resolution of the ADC, we expect $y[k] = x(kT_s) + q[k]$ where $q[k]$ is a discrete-time signal that represents the quantization error signal, also known as the quantization noise, of the ADC.

Quantization noise is generally assumed to be a discrete-time random sequence with a white spectrum that goes from DC to $f_s/2$, where $f_s = 1/T_s$.

One way to characterize an ADC and its accuracy in converting an analog to a digital signal is by measuring its output signal-to-quantization-noise ratio (SQNR), defined as:

$$\text{SQNR} = 10\log\left(\frac{\text{signal power}}{\text{quantization noise power}}\right) = 20\log\left(\frac{\sigma_x}{\sigma_q}\right) \tag{6.30}$$

where σ_x and σ_q represent the standard deviations of the input signal and the quantization noise, respectively. SQNR is directly related to the ADC resolution (N), which is the

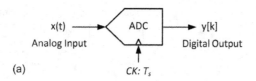

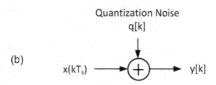

Figure 6.7 (a) ADC symbol, (b) a simple model of an ADC.

number of bits in the digital output of the ADC. Assuming a sinusoidal input, and a uniformly distributed quantization noise, it can be shown that:

$$\text{SQNR} = 6.02N + 1.76 \; [\text{dB}]. \tag{6.31}$$

This equation states that, given the same full-scale input, the higher the N, the lower the quantization noise power and the higher the SQNR.

For example, a 10-bit ADC is expected to have an SQNR of around 62dB whereas a 14-bit ADC is expected to have an SQNR of around 88dB. Also, note that every additional bit of resolution adds 6dB to the SQNR.

6.2.2 ADC Timing Error

The finite resolution (N) of an ADC is only one contributor to its noise power at the output. Another important contributor is the error caused by the jitter in the sampling clock [77], referred to as the timing error. Figure 6.8 models the ADC as a sampler followed by a quantizer. Even though in reality the functions of sampling and quantizing occur within a single block, separating the two is helpful in distinguishing the two contributors to the noise at the output. The sampler is expected to sample the input data at T_s intervals, but due to jitter, the samples are taken at $kT_s + \mathbf{a}_k$ where $\mathbf{a}_k$ represents the absolute jitter in the clock edge. As a result, the analog sample $s(k)$ that is fed to the quantizer is different from the intended input sample $x(kT_s)$ by $e[k]$. The quantizer, in turn, adds quantization error to $s[k]$ to produce the digital output, $y[k]$. Since these two errors, $e[k]$ and $q[k]$, in our model simply add to the output signal, there is no way to distinguish between the two; they both appear as noise at the output and degrade the SNR (we use SNR to refer to the ratio of signal to the total noise, including the quantization noise and the jitter noise, that is the output noise due to clock jitter). For this reason, it is common to define the effective number of bits (ENOB) of an ADC as follows:

$$\text{ENOB} = \frac{\text{SNR} - 1.76}{6.02}. \tag{6.32}$$

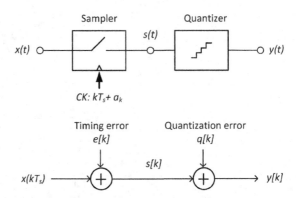

Figure 6.8 ADC is modeled as a sampler followed by a quantizer.

Note that, in general, the noise power comes from thermal noise, jitter noise, non-linearity, and quantization noise. If we ignore thermal noise, jitter noise, and the nonlinearity, then SNR = SQNR, and the above equation simply produces ENOB = N, that is, the nominal resolution of the ADC. On the other hand, in the presence of jitter, thermal noise, and nonlinearity, the total noise will be higher than the quantization noise, and therefore, SNR < SQNR, which results in an ENOB less than N. Clearly, a higher jitter will increase the total noise and therefore reduces ENOB. We are interested in quantifying the effect of jitter on ENOB. In particular, we are interested in knowing the jitter level at which the jitter noise becomes equal to the quantization noise. Also, we are interested in knowing what jitter levels will have no significant impact on ENOB.

Analysis: Based on the simplified model shown in Figure 6.8, the ADC output can be written as:

$$y[k] = x(kT_s) + e[k] + q[k] \tag{6.33}$$

where $e[k]$ and $q[k]$ are the jitter noise and the quantization noise, respectively. In this simplified model, we have assumed that the jitter noise and the quantization noise are independent of each other, and simply add up at the output of the ADC. Accordingly, the total noise power at the output of the ADC is equal to the sum of their powers. Let us now derive an expression for the jitter noise power while assuming $q[k] = 0$ (i.e., assuming an infinite resolution for the quantizer). Referring to Figure 6.9, we can approximate the output value at the k-th clock edge, $y[k]$, based on the Taylor expansion of the input signal, $x(t)$, at $t = nT_s$:

$$y[k] = x(kT_s + \mathbf{a}_k) = x(kT_s) + \left.\frac{dx}{dt}\right|_{t=kT_s} \mathbf{a}_k \tag{6.34}$$

where $\mathbf{a}_n$ represents the absolute jitter sequence at the n-th clock edge, and the higher-order terms in the Taylor expansion are assumed to be negligible compared to the first-order term. Accordingly, the error in the output due to jitter can be written as:

$$e[k] = \left.\frac{dx}{dt}\right|_{t=kT_s} \mathbf{a}_k. \tag{6.35}$$

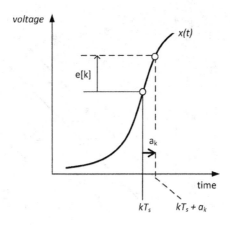

Figure 6.9 Jitter in a sampling clock contributes to the output quantization noise.

Note that, in general, both $x(t)$ and $\mathbf{a}_k$ are random processes, and so are dx/dt and $e[k]$. This equation simply states that the jitter is multiplied by the signal derivative to contribute to the total noise. The higher the signal derivative, the greater the impact of jitter on the total noise and the higher the deterioration of the SNR. For example, jitter will have little or no impact if the input is a DC or low-frequency signal. But the same jitter will have a strong impact on signals with higher frequencies. To show this mathematically, let us first consider the case where the input is a sinusoid with a full-scale amplitude A and a random uniformly distributed phase ϕ. In other words, $x(t) = A \sin(\omega_0 t + \phi)$. The assumption of *random phase* makes sense because we assume no phase relationship between the input and the sampling clock. One can easily verify that:

$$E[x(t)] = E[dx/dt] = 0 \tag{6.36}$$

$$\sigma_x^2 = E[x^2(t)] = A^2/2 \tag{6.37}$$

$$\sigma_{dx/dt}^2 = E[(dx/dt)^2] = A^2\omega_0^2/2. \tag{6.38}$$

Hence, we can write (if we assume dx/dt and $\mathbf{a}_k$ are uncorrelated):

$$\sigma_e^2 = E[e^2[k]] = \frac{A^2\omega_0^2}{2}\sigma_{\mathbf{a}}^2 \tag{6.39}$$

where $\sigma_{\mathbf{a}}^2$ represents the jitter variance and σ_e^2 is the noise power due to jitter. It is common to write the noise power in dBFS, which is defined as the noise power in dB relative to the power of the full-scale signal ($A^2/2$). Therefore, we can write:

$$\sigma_e^2 = 10 \log(\frac{A^2\omega_0^2\sigma_{\mathbf{a}}^2/2}{A^2/2}) = 20 \log(2\pi f_0 \sigma_{\mathbf{a}}) \ \text{[dBFS]} \tag{6.40}$$

In contrast, the quantization noise power can be written in dBFS as follows:

$$\sigma_q^2 = \frac{V_{LSB}^2/12}{A^2/2} = -1.76 - 6.02N \ \text{[dBFS]} \tag{6.41}$$

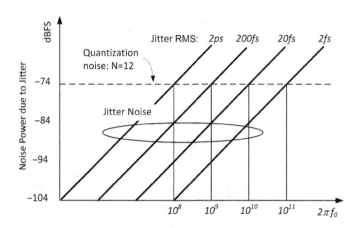

Figure 6.10 Noise power due to jitter as a function of signal frequency.

where V_{LSB} represents the change in voltage that corresponds to the least-significant bit (LSB), and the peak-to-peak amplitude of the signal, $2A$, represents the ADC full-scale input, i.e., $2A = 2^N V_{LSB}$.

Equations 6.40 and 6.41 show that, while the quantization noise power is not a function of the input frequency, the jitter noise power increases with frequency. This is plotted in Figure 6.10 for an ADC with $N = 12$. The quantization noise in this case is −74dBFS. The jitter noise, however, increases with both the RMS jitter and the input signal frequency. For a fixed RMS jitter, such as $\sigma_a = 2$ps, the jitter noise will increase with frequency until it is equal to the quantization noise at $\omega_0 = 100$Mrad/s. If the signal frequency increases beyond this value, then the output noise power will be dominated by jitter, and not by the quantization noise. Similarly, for a fixed input frequency, such as 1Grad/s, the jitter noise increases linearly with RMS jitter. At $\sigma_a = 0.2$ps, the jitter noise will be equal to the quantization noise. Any increase in the jitter RMS will make the jitter noise dominant in this design.

Example: A 10-bit ADC is designed to work with a sampling clock with an RMS jitter of 1ps. (a) Find the input frequency at which the quantization noise power equals the jitter noise. (b) Calculate *ENOB* at this frequency. (c) Find the input frequency at which the jitter noise power is 10% of the quantization noise power.

Solution: (a) For $N = 10$, using Equations 6.41 and 6.40, and equating σ_q^2 and σ_e^2, we will have $-61.96 = 20\log(2\pi f_0 \times 10^{-12})$. Hence $f_0 = 127$MHz. (b) At this frequency, the total noise power will be twice the quantization noise power, i.e., 3dB higher than −61.96 [dBFS], or −58.96 [dBFS]. Using Equation 6.32, we will have *ENOB* = 9.5. (c) 10% is equivalent to 10dB. Therefore, the jitter noise will be −71.96[dBFS]. Using Equation 6.40, we will have $f_0 = 40.2$MHz.

We now consider the general case where $x(t)$ is a random process in Equation 6.34, and derive an expression for the output noise power due to sampling clock jitter. Taking the autocorrelation function of both sides of Equation 6.34 results in:

$$R_y(mT_s) = R_x(mT_s) + R_e(mT_s). \tag{6.42}$$

It can also be shown [77] that:

$$R_e(mT_s) = -R_x''(mT_s)\, R_{\mathbf{a}}(mT_s). \tag{6.43}$$

Evaluating this equation at $m = 0$ results in:

$$R_e(0) = -R_x''(0)\, R_{\mathbf{a}}(0). \tag{6.44}$$

Recognizing that $R_e(0) = \sigma_e^2$ and $R_{\mathbf{a}}(0) = \sigma_{\mathbf{a}}^2$, we can write:

$$\sigma_e^2 = -R_x''(0)\, \sigma_{\mathbf{a}}^2. \tag{6.45}$$

This equation states that the output noise power is the product of the jitter power and a constant that depends only on the input signal autocorrelation around zero. This constant, for the case of a sinusoidal input with a random phase, turns out to be $A^2\omega_0^2/2$, as found in Equation 6.39. Also note that the jitter noise power is independent of the shape of the jitter spectral density; instead, it is a function of the total jitter power $\sigma_{\mathbf{a}}^2$ [77].

For further insight, let us take the Fourier transform of both sides of Equation 6.43. We will have:

$$S_e(f) = [(2\pi f)^2 S_x(f)] * S_{\mathbf{a}}(f) \tag{6.46}$$

where $*$ denotes convolution, and $S_e(f)$, $S_x(f)$, and $S_{\mathbf{a}}(f)$ represent the power spectral densities of the error signal due to jitter, the input signal, and the clock jitter, respectively. This equation clearly shows that the higher frequencies contribute more to noise power than the low frequencies due to the factor $(2\pi f)^2$ in the equation.

6.2.3 Design Considerations

In designing ADCs, depending on the availability of a low-jitter clock (from a PLL, for example), and the target resolution, it is easy to find out if the SNR is limited by the quantization noise or by the jitter. Given Equations 6.40 and 6.41, it is easy to compare the two noise powers. If the total SNR design is limited by the quantization noise – that is, when the quantization noise is an order of magnitude larger than the jitter noise – then reducing the clock jitter will not improve the SNR. If, on the other hand, the total SNR is limited by the jitter noise – that is, when the jitter noise is comparable to or higher than the quantization noise – an effort should be made to improve the clock jitter. If the jitter noise is several orders of magnitude higher than the quantization noise and cannot be improved, then perhaps the design could use a smaller resolution (N) without sacrificing the overall SNR. This should reduce the ADC power consumption.

6.3 Effects of Timing Skew in Time-Interleaved ADCs

To accommodate sampling rates beyond what a typical ADC can handle, we can sample the input signal with N ADCs (also called sub-ADCs) by using N phases of the clock in an architecture known as N-way time-interleaved (T/I) ADC. This architecture increases the sampling rate by a factor of N at the cost of increasing the area by N while maintaining a good power efficiency. As an example, to build an ADC that could sample the input at 10GS/s, [78] uses four time-interleaved flash ADCs where each ADC samples the input at 2.5GS/s. To accommodate 90GS/s with 8-bit resolution, [79] uses 64 T/I SAR ADCs, each operating at 1.4GS/s. In this section, we study the effect of timing skew on a general N-way T/I ADCs, where N is typically less than 64.

6.3.1   Background

Figure 6.11 shows a block diagram of a four-way T/I ADC, which consists of four flash ADCs, each clocked with one of four phases of the clock. Due to interleaving, each sub-ADC samples the input at $f_s/4$ for an aggregate sampling frequency of f_s. Under ideal conditions, i.e., where the four phases are exactly $T_s = T/4$ apart, it may not be possible to distinguish between this ADC and an ADC that samples the input directly at f_s, i.e., without interleaving. However, as we will see next, any skew among the clock phases will introduce error to the output, degrading its SQNR.

6.3.2   Effects of Timing Skew

There are several non-idealities that can impact the performance of a T/I ADC. These include sub-ADC offsets and gain errors, which may be different among the sub-ADCs, and timing skew among the clock phases. We concentrate here only on the effects of timing skew on performance degradation of the ADCs.

In an ideal N-way T/I ADC, the N phases of the clock are expected to be exactly T_s apart in time. However, due to timing skew caused mainly by layout asymmetries, the N phases usually deviate from their ideal times, say by τ_0 to τ_{N-1}. These deviations are usually constant over time and hence they are referred to as *skew*, not jitter. Nevertheless, their impact on the ADC output will appear as high-frequency jitter, as we will see next.

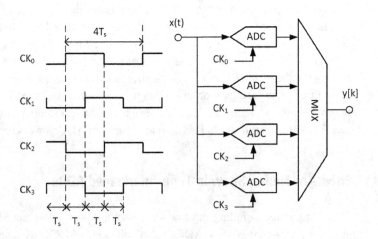

Figure 6.11 Block diagram of a 4-way time-interleaved ADC.

Assume the input of the ADC is a stationary process denoted by $x(t)$. Under ideal sampling conditions, i.e., no skew, the output of the ADC can be represented as $x(kT_s + \hat{\tau})$, where $\hat{\tau}$ represents a fixed delay for all samples. Due to skew, however, the output of the ADC will be of the form $y[k] = x(kT_s + \tau_k)$ where $\tau_k = \tau_{(k \mod N)} \in \{\tau_0, \tau_1, \ldots, \tau_{N-1}\}$. As a result, the error due to skew can be written as:

$$e[k] = x(kT_s + \tau_k) - x(kT_s + \hat{\tau}). \tag{6.47}$$

Note that $e[k]$ is a cyclostationary process whose properties repeat every N cycles. This makes intuitive sense because $e[k]$ and $e[k + N]$ correspond to the output of the same ADC in the N-way T/I system, and hence have the same statistical properties. To find the average power of $e[k]$, we first find the average power of $e[k]$ over N consecutive cycles to create a new stationary process denoted by $\overline{e^2}$:

$$\overline{e^2} = \frac{1}{N} \sum_{k=0}^{N-1} e^2[k] = \frac{1}{N} \sum_{k=0}^{N-1} (x(kT_s + \tau_k) - x(kT_s + \hat{\tau}))^2. \tag{6.48}$$

We now determine the expected value of this stationary process as follows:

$$E[\overline{e^2}] = \frac{1}{N} \sum_{k=0}^{N-1} E[x^2(kT_s + \tau_k) + x^2(kT_s + \hat{\tau}) - 2x(kT_s + \tau_k)x(kT_s + \hat{\tau})]. \tag{6.49}$$

The first two terms in this equation are equal to $R_x(0)$ since $x(t)$ is stationary, while the last term is equal to $2R_x(\tau_k - \hat{\tau})$. Therefore, we can write:

$$E[\overline{e^2}] = 2R_x(0) - \frac{2}{N} \sum_{k=0}^{N-1} R_x(\tau_k - \hat{\tau}). \tag{6.50}$$

This equation relates the power of error due to skew to the statistical properties of the input signal $x(t)$ and to the skew sequence τ_k. Since the input signal power is $R_x(0)$, we can write an expression for SNR due to skew alone as follows:

$$SNR_\tau = \frac{R_x(0)}{2(R_x(0) - \frac{1}{N} \sum_{k=0}^{N-1} R_x(\tau_k - \hat{\tau}))}. \tag{6.51}$$

The error power, as found in Equation 6.50, is a function of $\hat{\tau}$. This power can be minimized if we set its derivative with respect to $\hat{\tau}$ to zero. In other words, to minimize $E[\overline{e^2}]$, we must have:

$$\frac{\partial}{\partial \hat{\tau}} \sum_{k=0}^{N-1} R_x(\tau_k - \hat{\tau}) = 0. \tag{6.52}$$

To solve this equation for $\hat{\tau}$, we expand $R_x(\tau_k - \hat{\tau})$ using the first three terms in the Taylor series:

$$R_x(\tau_k - \hat{\tau}) \approx R_x(0) + R'_x(0)(\tau_k - \hat{\tau}) + \frac{R''_x(0)}{2}(\tau_k - \hat{\tau})^2. \tag{6.53}$$

Using this equation and assuming $R'_x(0) = 0$ (we will return to this assumption later), we can rewrite Equation 6.52 as follows:

$$\sum_{k=0}^{N-1} R_x''(0)(\tau_k - \hat{\tau}) = 0 \tag{6.54}$$

and hence:

$$\hat{\tau} = \frac{1}{N} \sum_{0}^{N-1} \tau_k. \tag{6.55}$$

This equation simply states that the minimum error power occurs when the ideal ADC delays the input signal by the average of all the skews prior to sampling them. The resulting SNR_τ in this case is still obtained from Equation 6.51 with the value of $\hat{\tau}$ calculated from Equation 6.55.

To provide further insight into the expression for SNR_τ, let us substitute $R_x(\cdot)$ in Equation 6.51 with its Taylor expansion in Equation 6.53. After some manipulation, we can write:

$$SNR_\tau = \frac{R_x(0)}{-R_x''(0)\frac{1}{N}\sum_{k=0}^{N-1}(\tau_k - \hat{\tau})^2} = \frac{R_x(0)}{-R_x''(0)\sigma_\tau^2} \tag{6.56}$$

where σ_τ^2 represents the skew variance. This equation shows that the noise power due to skew is given by $-R_x''(0)\sigma_\tau^2$, which is identical in form to the noise power due to jitter as found in Equation 6.40 for Nyquist ADCs.

In deriving Equation 6.51, we assumed a scaling factor of unity for the output of the an ideal ADC. It has been shown [80], however, that an optimal scaling factor, which may be different than 1, due to gain mismatch for example, will result in the following expression for SNR_τ:

$$SNR_\tau = \frac{R_x(0)}{R_x(0) - \frac{1}{N}\sum_{k=0}^{N-1}R_x(\tau_k - \hat{\tau})} \cdot \frac{R_x(0)}{R_x(0) + \frac{1}{N}\sum_{k=0}^{N-1}R_x(\tau_k - \hat{\tau})}. \tag{6.57}$$

This expression reduces to the one in Equation 6.51 if we approximate $\frac{1}{N}\sum_{k=0}^{N-1}R_x(\tau_k - \hat{\tau})$ by $R_x(0)$ in the second fraction in Equation 6.57. This approximation makes sense if $R_x(\tau)$ is flat around 0; in other words, if $R_x'(0) = 0$. When this is not true for $R_x(\tau)$, we must resort to Equation 6.57 without any approximation.

6.3.3   Alternative Approach

When the input to the ADC is a sinusoid with a random phase, the expression for the SNR_τ can be found more intuitively. Let us assume the input to the ADC is a sinusoid with random phase, $x(t) = A \sin(\omega_0 t + \phi)$. Then the error signal caused by skew can be written as:

$$e[k] = dx/dt|_{kT_s}, \ \tau_k = \tau_k A\omega_0 \cos(\omega_0 kT_s + \phi)) \tag{6.58}$$

where $e[k]$ is a product of two random processes: τ_k and $\cos(\omega_0 kT_s + \phi)$. Without a loss of generality, we can assume $E[\tau_k] = 0$. If we also assume the two processes are uncorrelated, we can write:

$$\sigma_e^2 = A^2\omega_0^2 E[\tau_k^2]E[\cos^2(\omega_0 kT_s + \phi)] = A^2\omega_0^2\sigma_\tau^2/2 \tag{6.59}$$

where σ_τ^2 represents the variance of the timing skew among all the clock phases. If we express this power relative to the power of the input signal, we will have:

$$\sigma_e^2 = 10 \log \left(\frac{A^2 \omega_0^2 \sigma_\tau^2 / 2}{A^2/2} \right) = 20 \log \left(2\pi f_0 \sigma_\tau \right) \quad \text{[dBFS]}. \qquad (6.60)$$

This equation is identical in form to Equations 6.60 and 6.40, implying that both clock jitter and clock skew have similar impacts on noise power and in limiting the SNR. In other words, a clock jitter and a clock skew with the same RMS will produce the same noise level [80], [81], [82].

6.3.4   Design Considerations

Unlike clock jitter, clock skews are generally constant over time, and hence it is possible to correct them or calibrate them. The effects of clock skew can be corrected by digital signal processing following the sub-ADCs, as shown in Figure 6.12. In this scheme, the clock skews are left uncorrected, but their effects are removed through adaptive filters. The filters are often fractionally spaced so as to provide interpolation between the sampled values in the digital domain. However, this approach can be power- and area-hungry, as the filters often require several taps. For example, [83] uses a 33-tap FIR filter for skew correction of a 12-way T/I SAR ADC, operating at 1.62GS/s.

In contrast to this approach, the effects of skew can be observed, for example at the output of the sub-ADCs, in order to calibrate the skew by analog means. This approach, as shown in Figure 6.13, attempts to minimize the clock skew (σ_τ) through a control loop.

One example of this approach is shown in Figure 6.14 for a 10GS/s 4-way T/I ADC, used in a 10Gb/s wireline receiver application. In this example, the input to the receiver is a 10Gb/s random binary signal that is subjected to frequency-dependent attenuation by the channel. The attenuated signal needs to be sampled, converted to digital, and equalized to compensate for the channel attenuation. However, due to the speed limits of

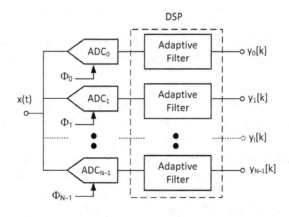

Figure 6.12 Removing the effects of clock skew with adaptive filters.

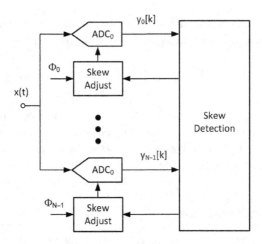

Figure 6.13 Block diagram implementation of a skew detection and correction technique.

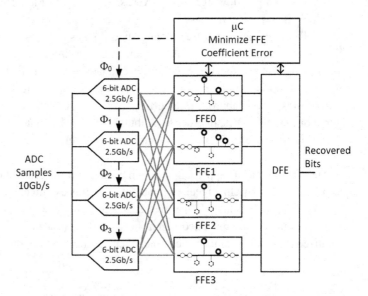

Figure 6.14 Skew calibration using FFE outputs for a four-way time-interleaved ADC. © 2013 IEEE. Reprinted, with permission, from [78].

the process, the cascade of ADC and feed-forward equalizer (FFE) is time-interleaved to work at 2.5GS/s. The time-interleaved FFEs are four replicas of the full-rate FFE, except that they work on 1/4 of the full data rate. As such, the FFEs are expected to have identical coefficients to those of a full-rate FFE. However, due to the skew in the four phases of the clock, the four ADC outputs will contain errors, and this will force the optimum coefficient values to be different among the four FFEs. For example, assume that the first phase of the clock is slightly off from its ideal time while all other phases are at their ideal times. In other words, assume $\tau_0 \neq 0$, $\tau_1 = \tau_2 = \tau_3 = 0$. To compensate for this error, the main tap of FFE0 should be changed slightly while its remaining taps should be at their original values since other ADCs have no errors. In FFE1, FFE2, and FFE3, the second, third, and fourth taps should be changed, respectively, as they correspond to the inputs from ADC0, while all their other taps should remain at their original values. An adaptive engine in this design monitors the filter coefficients and attempts to minimize their spread by adjusting the skew of the clock phases.

6.4   Effects of Jitter on Continuous-Time $\Delta\Sigma$ Modulators

6.4.1   Background

A Nyquist ADC samples its input at twice the input bandwidth so as to avoid aliasing. In contrast, an oversampling ADC samples its input at a rate that is several times higher than twice the bandwidth. If we assume the input bandwidth is f_0 and the sampling frequency is f_s, then we define the oversampling ratio (OSR) as follows:

$$OSR = \frac{f_s}{2f_0}. \tag{6.61}$$

Accordingly, the Nyquist rate (i.e., $f_s = 2f_0$), corresponds to $OSR = 1$, whereas a typical OSR in an oversampling ADC could be 8, 16, or higher. We will review the basics of an oversampling continuous-time $\Delta\Sigma$ ADC here, and analyze its performance degradation due to jitter. [84] provides a complete background review of this topic.

Figure 6.15 shows a block diagram of a first-order continuous-time $\Delta\Sigma$ modulator (CT-DSM) along with its linear model. The modulator consists of a feedback loop that includes a flash ADC, or simply a quantizer, and a DAC, both clocked at f_s, and a continuous-time integrating filter. As we mentioned in relation to Figure 6.8, the flash ADC can be modeled as a summer that adds to the input both the quantization error, $q[k]$, and the error due to non-ideal timing such as jitter, $e_F[k]$. As we saw in Section 6.1, the DAC can also be modeled as a summer that adds to its input the static error $e_S[k]$ due to mismatch among its elements, and the dynamic errors due to timing $e_D[k]$. Since the effects of timing errors in the ADC and the DAC are captured by their respective error sources, the switch in Figure 6.15(b) is assumed to operate at f_s with no timing error.

The impulse response of the DAC in Figure 6.15(a) is represented by $h_D(t)$ in Figure 6.15(b). For an NRZ DAC, $h_D(t) = u(t) - u(t - T_s)$, and its corresponding Laplace transform will be $H_D(s) = (1 - e^{-sT_s})/s$. Note that in the model of Figure 6.15(b), there is only one signal source, which is $x(t)$, while there are four sources of error: $q[n]$, $e_F[n]$, $e_S[n]$, and $e_D[n]$. For a basic calculation of the output SQNR, let us first ignore all the error sources except for the quantization noise of the ADC, $q[n]$. In this case, the signal transfer function (STF) and the noise transfer function (NTF), can be expressed as follows:

$$STF = H_D(s)z^{-1} = \frac{1 - e^{-sT_s}}{s}e^{-sT_s} \tag{6.62}$$

$$NTF = 1 - z^{-1} = 1 - e^{-sT_s} \tag{6.63}$$

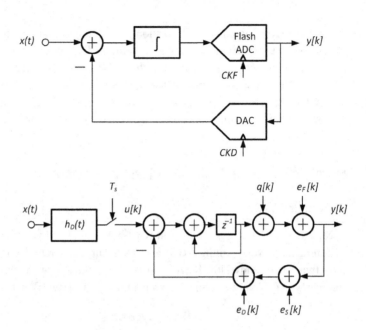

Figure 6.15 (a) Block diagram of a continuous-time $\Delta\Sigma$ modulator, (b) linear model of the modulator.

The power spectral density of the output, $S_y(f)$, then can be written as:

$$S_y(f) = S_x(f) \, |STF|^2 + S_q(f) \, |NTF|^2 \tag{6.64}$$

where $S_x(f)$ and $S_q(f)$ represent the power spectral densities of the input and the quantization error, respectively. In the signal band of interest (0 to f_0), $|STF(j\omega)|^2 \approx 1$ and $|NTF(j\omega)|^2 \approx \omega^2$. Since the total quantization noise power is constant at $V_{LSB}^2/12$ (see Equation 6.41), the frequency-dependent NTF allocates less of this power to the signal band and more to the out-of-band frequencies, as depicted in Figure 6.16. As a result, the input signal arrives at the output unaltered whereas the quantization noise arrives attenuated in the signal band. This is the major advantage of this modulator over the Nyquist ADC as the loop shapes (filters) the quantization noise before contributing to the total noise level at the output. It can be shown [84] that SQNR of a first-order modulator for a sinusoidal input can be written as:

$$SQNR = \frac{\int_0^{f_0} S_x(f) \, |STF|^2 df}{\int_0^{f_0} S_q(f) \, |NTF|^2 df}$$
$$= 6.02N + 1.76 - 5.17 + 30\log(OSR). \tag{6.65}$$

This equation states that the $SQNR$ of an oversampling ADC improves by 6dB for every extra bit of resolution and 9dB for every doubling of OSR. In other words, every doubling of OSR is equivalent to adding 1.5 bits of resolution to the ADC.

It can be shown [84] that for a second- and third-order DSM, the SQNR improves by 15dB (equivalent to 2.5 bits) and 21dB (equivalent to 3.5 bits), respectively, for every doubling of OSR.

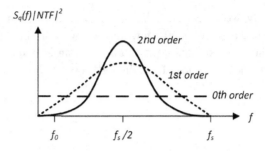

Figure 6.16 Power spectral density of quantization noise at the output of a $\Delta\Sigma$ modulator.

6.4.2 Effects of Flash ADC Timing Error on SNR

Figure 6.15(b) models the timing error associated with the flash ADC as $e_F[k]$. We are interested in finding the degradation in SNR associated with this timing error. For this purpose, we assume all other error sources, including the quantization error, is zero.

The power spectral density of the output can be written in terms of the power spectral densities of the input and the flash ADC timing error as follows:

$$S_y(f) = S_x(f) |STF|^2 + S_{eF}(f) |NTF|^2 \tag{6.66}$$

where $S_{eF}(f)$ represents the PSD of the flash ADC's timing error under open-loop conditions. This PSD is filtered by the $|NTF|^2$, similar to quantization noise, as in Equation 6.64. Accordingly, the total noise at the output of the modulator (due to timing error only) will be:

$$Output\ Noise\ Power = \int_0^{f_0} S_{eF}(f) |NTF|^2 df \tag{6.67}$$

In general, $S_{eF}(f)$ takes the form of Equation 6.46. However, if we assume $S_{eF}(f)$ to be flat in the 0 to f_s region, then the output noise power is suppressed by first limiting the in-band noise to $1/OSR$ of the corresponding Nyquist ADC, and second, by shaping the PSD of the in-band noise by $|NTF|^2$. The latter will render this noise insignificant compared to the corresponding noise from the DAC timing error as we will see next.

6.4.3 Effects of DAC Timing Error on SNR

The DAC timing error is modeled as $e_D[k]$ in Figure 6.15(b). Following a similar procedure to the previous section, we can write an expression for the power spectral density of the output in terms of the power spectral densities of the input and the DAC timing error:

$$S_y(f) = S_x(f) |STF|^2 + S_{eD}(f) |STF|^2. \tag{6.68}$$

This equation shows that the DAC timing error is shaped by $|STF|^2$, which is expected to be unity in the signal band. This is in contrast with the flash ADC timing error that is heavily attenuated by the NTF in the signal band. It is for this reason that the SNR of a $\Delta\Sigma$ modulator is often determined by the timing error of its DAC, not of its flash ADC.

Finally, we expect the noise due to DAC timing error to have an average power as derived in Equation 6.20. However, we note that in the case of $\Delta\Sigma$ modulators, the DAC input is expected to have a higher transition density (σ_Δ), and this will further increase the noise due to DAC timing errors.

7 Effects of Jitter in Wireline Applications

A wireline communication system sends and receives data based on precise timing. Any deviation from the ideal timing, i.e., jitter, has a direct impact on the bit error rate (BER) at the receiver. Figure 7.1 shows a generic block diagram of a wireline transceiver that consists of a transmitter, a channel, and a receiver. The role of the transmitter is to serialize N binary sequences, generated by the core logic, into one binary sequence D_{TX}, and send it reliably over the channel to the receiver. If we assume, for example, that the core logic produces 16 data sequences (i.e., $N = 16$), each at 500Mb/s, then the transmitter will produce a single binary sequence at 8Gb/s. The channel could be a few meters of cable (e.g., between a computer and a peripheral), a few inches of PCB trace (e.g., between a microprocessor and a memory module on the same board), or a few millimeters of an interconnect in a multi-chip module. In all these cases, the channel tends to attenuate the high-frequency components of the signal, and degrade the received eye both in the voltage domain (vertical axis) and in the time domain (horizontal axis). The received signal timing is further degraded due to the jitter in the transmitter clock (CK_{TX}), which in turn may be caused by thermal noise in the clock generation circuit or by power supply noise. In the presence of all these jitter components, the receiver must be designed so as to recover the transmit bits reliably, e.g., at a BER better than 10^{-12} or 10^{-15}. In this chapter, we first review various sources of the jitter that limits the BER, and then provide techniques for jitter characterization, jitter monitoring, and jitter mitigation.

As mentioned in Chapter 1, jitter does not always work against performance; there are cases where we can use jitter to improve system performance. We provide two such examples later in this chapter.

7.1 Basic Concepts in Wireline Signaling

A wireline transmitter, in its simplest form, sends a non-return-to-zero (NRZ) binary signal to the receiver. The wireline receiver first equalizes the received signal so as to compensate for the channel loss (or equivalently to remove the channel ISI) and then samples the equalized signal, using a recovered clock at the center of the data eye to see whether the sampled value is above or below a threshold. If the sample is above the threshold, the data bit is considered a 1, and if below the threshold, it is considered a 0. Let us assume for simplicity that the equalized data at the receiver is ideal, i.e., it has no

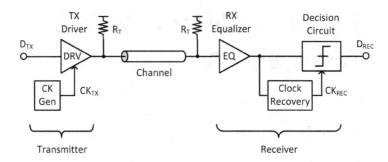

Figure 7.1 Block diagram of a wireline transceiver.

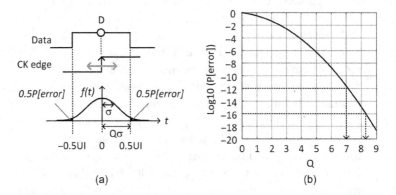

Figure 7.2 (a) Gaussian jitter in clock edge results in data decision error, (b) probability of error as a function of Q.

jitter, but that the recovered clock has some jitter. We are interested in calculating the probability of error in data decisions. In other words, we are interested in estimating the BER of this system.

Figure 7.2(a) shows a sketch of the equalized data at the receiver, where a lone 1 appears among neighboring zeros. Clearly if we sample this data bit in the center of the bit as shown, we will have no error. However, assume the clock edge sampling the data contains Gaussian jitter as depicted in the figure. Since the jitter is Gaussian, it is unbounded, and, therefore, there is a probability, however small, that the clock edge may move by more than 0.5UI either to the right or to the left so as to miss sampling the current bit. If either of the two events occurs, we will record a 0 instead of a 1, causing an error.

The probability of error is equal to the shaded area under the tails of the Gaussian curve. As we will see in Section 9.2, this probability will depend on Q, which is defined as the ratio of 0.5UI, in this case, to the jitter standard deviation, σ. Accordingly, we can write:

$$P\,[\text{error}] = \text{erfc}(\frac{Q}{\sqrt{2}}) \qquad (7.1)$$

where erfc(x) is the complementary error function, as defined in Equation 9.16.

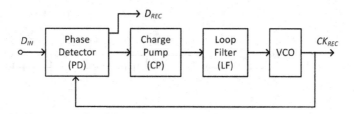

Figure 7.3 Block diagram of clock and data recovery (CDR).

A graph of the error probability versus Q is shown in Figure 7.2(b). The probability of error decreases exponentially as Q is increased, corresponding either to a reduction in the standard deviation of the jitter or to a reduction in the data rate. For example, a transceiver at 2.5Gb/s may exhibit a BER of 10^{-12}. If we increase the data rate to 5Gb/s (assuming all the components work at 5Gb/s, and assuming jitter remains constant irrespective of the data rate), Q will decreases by a factor of 2 (because 1UI is now half as long) and this will increase the BER by several orders of magnitude to 10^{-4}.

7.2 Jitter in Analog CDR

Figure 7.3 shows a block diagram of an analog clock and data recovery (CDR) unit that consists of a phase detector (PD), a charge pump (CP), a loop filter (LF), and a voltage-controlled oscillator (VCO), all in a single loop. This block diagram is equivalent of the combined clock recovery and decision circuit blocks in Figure 7.1, with the decision circuit now embedded in the PD block. The PD has two inputs: the equalized data D_{IN} and the recovered clock CK_{REC}, and one output which represents the phase error between the two input signals. The PD output is then integrated (via the CP and LF) and fed as a control voltage to the VCO so as to influence its phase and frequency. When the CDR reaches steady state, i.e., when the average PD output becomes zero, the VCO frequency stays constant. The VCO output forms the recovered clock CK_{REC} which can be used to retime the equalized data, using a flip-flop, to produce the recovered data D_{REC}.

The VCO in the CDR may be a ring or an LC oscillator, but in either case its output contains random jitter. Therefore, when we sample and retime the equalized data, we are retiming it with a jittery clock, although the jitter in this clock is shaped by the feedback loop, as we see later in this section.

Let us now create an equivalent block diagram of the CDR with a focus on jitter.

7.2.1 Linear Model of the CDR

Figure 7.4 shows a linear model of the CDR in a *frequency lock* condition, that is, when the recovered clock frequency is equal to the data rate (in a full-rate system), but it has yet to achieve *phase lock*; that is, for the recovered clock phase to be driven towards the input data phase. Under these conditions, the clock frequency (and hence the data rate) is fixed, say, at f_0, and, as such it does not appear explicitly in the model, instead,

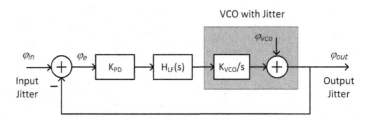

Figure 7.4 A linear model of a CDR takes data jitter as its input and provides the recovered clock jitter as its output.

the model is only concerned with the excess phase of the input data, denoted by φ_{in}, the phase of the VCO, denoted by φ_{VCO}, and the phase of the recovered clock φ_{out}. Since the frequency is fixed, these three phase variables, if expressed in terms of UI, also represent the jitter corresponding to the input data, the VCO, and the recovered clock, respectively. For example, a recovered clock phase instance of 0.4π radians is equivalent to a jitter instance of $0.4\pi/(2\pi f_0) = 0.2$UI. In other words, depending on whether we express φ_{out} in terms of radians or UI, we can view it as either excess phase or jitter. For simplicity, and as is common in wireline applications, we express all signals in terms of UI, and, as such, view φ_{in} as the input data jitter, φ_{VCO} as the VCO jitter, and φ_{out} as the recovered clock jitter. Also, we assume φ_{in} and φ_{VCO} are uncorrelated as they come from independent sources.

Using this model, the power spectral density (PSD) of jitter in the recovered clock $S_{out}(f)$ is a linear combination of the PSDs of the input data jitter $S_{in}(f)$ and the VCO jitter $S_{VCO}(f)$, shaped by two transfer functions. We can write:

$$S_{out}(f) = |H_T(f)|^2 S_{in}(f) + |H_G(f)|^2 S_{VCO}(f) \qquad (7.2)$$

where $|H_T(f)|^2$, known as *the jitter transfer function*, can be written as:

$$|H_T(f)|^2 = \left| \frac{K_{PD}K_{VCO}H_{LF}(f)}{j2\pi f + K_{PD}K_{VCO}H_{LF}(f)} \right|^2 \qquad (7.3)$$

and $|H_G(f)|^2$, known as *the jitter generation function*, can be written as:

$$|H_G(f)|^2 = \left| \frac{j2\pi f}{j2\pi f + K_{PD}K_{VCO}H_{LF}(f)} \right|^2. \qquad (7.4)$$

We will discuss these two functions and their significance in more detail in the following sections.

7.2.2   Jitter Transfer

Equation 7.2 shows that the PSD of the recovered clock jitter is a linear combination of the PSDs of the input jitter and the VCO jitter. In the absence of VCO jitter, i.e., when the VCO circuit does not contribute any jitter to its own output, $S_{VCO} = 0$, and hence the PSD of the recovered clock is the product of the PSD of the input jitter and $|H_T(f)|^2$. For convenience, we also refer to $H_T(f)$, not squared, as the *jitter transfer*

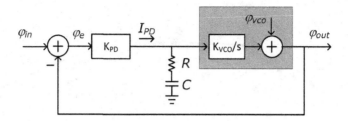

Figure 7.5 A linear model of a CDR with an RC loop filter.

function with the understanding that the ratio of the PSD of the output jitter to the PSD of the input jitter is $|H_T(f)|^2$. Note that $H_T(f)$ would represent the signal transfer function from the input to the output of CDR under the assumption that the input is a deterministic signal. For a random input signal, such as jitter, we typically use $|H_T(f)|$ when we deal with the jitter amplitude and $|H_T(f)|^2$ when we deal with the PSD of jitter.

We can illustrate a typical $H_T(f)$ by considering a CDR with a linear model, as shown in Figure 7.5.

Here, we have assumed the output quantity produced by the combined PD and CP (refer to Figure 7.3) is current and the loop filter is implemented using a series RC circuit with $H_{LF}(s) = R + 1/(sC)$. We can easily verify that

$$H_T(s) = \frac{\varphi_{out}(s)}{\varphi_{in}(s)}\bigg|_{\varphi_{VCO}=0} = \frac{K_{PD}K_{VCO}Rs + \frac{K_{PD}K_{VCO}}{C}}{s^2 + K_{PD}K_{VCO}Rs + \frac{K_{PD}K_{VCO}}{C}}. \tag{7.5}$$

If we define $K = K_{PD}K_{VCO}/C$ and $\tau = RC$, we can write:

$$H_T(s) = \frac{K(\tau s + 1)}{s^2 + K\tau s + K}. \tag{7.6}$$

Alternatively, to write this equation in a canonical form, we define:

$$\omega_n^2 := \frac{K_{PD}K_{VCO}}{C}$$

$$\xi := \frac{K_{PD}K_{VCO}R}{2\omega_n} \tag{7.7}$$

where ω_n and ξ are known as the natural frequency and the damping factor, respectively. Using this notation, we can rewrite Equation 7.5 as follows:

$$H_T(s) = \frac{2\xi\omega_n s + \omega_n^2}{s^2 + 2\xi\omega_n s + \omega_n^2}. \tag{7.8}$$

Replacing s with $j\omega$ ($= j2\pi f$), we can derive the jitter transfer function as a function of jitter frequency:

$$|H_T(f)|^2 = \frac{4\xi^2\omega_n^2\omega^2 + \omega_n^4}{(\omega_n^2 - \omega^2)^2 + 4\xi^2\omega_n^2\omega^2}. \tag{7.9}$$

A Bode plot of $|H_T(f)|$ is shown in Figure 7.6, corresponding to three different values of $\xi \in \{0.5, 1, 2\}$. In all three cases, the jitter transfer represents a low-pass filter whose

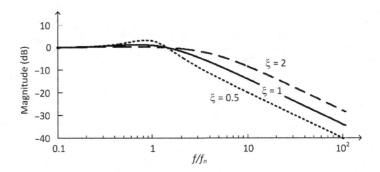

Figure 7.6 Jitter transfer function for three different values of ξ.

magnitude is around 1 (0 dB) for low jitter frequencies and drops at 20 dB/decade for frequencies above ω_n. In other words:

$$|H_T(f)|^2 = \begin{cases} 1 & \text{if } f << f_n \\ 4\xi^2(\frac{f_n}{f})^2 & \text{if } f >> f_n. \end{cases} \qquad (7.10)$$

In simple terms, the first line of Equation 7.10 says the recovered clock *tracks* the low-frequency jitter of the input data. In other words, if the data transitions move slowly around their nominal locations, the CDR control loop would be able to track these movements such that the error between the data and clock edges would be zero (on average). This is, in fact, the characteristic of any control loop with at least one integration in the loop. The second line of this equation shows that the jitter transfer magnitude is less than unity, meaning that the recovered clock does not track the high-frequency data jitter as well as it tracks low-frequency data jitter. At the limit, the jitter transfer from input to the recovered clock is zero, and the recovered clock does not track the data jitter at all. This means that the recovered clock does not suffer from high-frequency jitter even though the input signal may contain high-frequency jitter. As we will discuss later, this characteristic of the recovered clock will limit the CDR tolerance to high-frequency jitter.

Jitter Peaking in Jitter Transfer Function

The jitter transfer function, as shown in Figure 7.7, has a peak that is slightly larger than 1 (0dB). This peaking could be anywhere from 0.07dB to 0.4dB when we vary ξ from 5 to 2. The peaking implies that jitter will be amplified at some frequencies in the CDR, producing a jitter amplitude in the recovered clock, and thus also in the recovered data, that is slightly larger than the jitter amplitude in the input data. This is certainly undesirable, especially in applications such as repeaters where the recovered data from one CDR may be launched onto a channel and subsequently become the input of another CDR. In this case, the jitter will keep increasing following each repeater, contributing to an increase in BER. For this reason, we wish to reduce, or possibly eliminate, the jitter peaking. First, let us examine what causes this peaking and then how to reduce it.

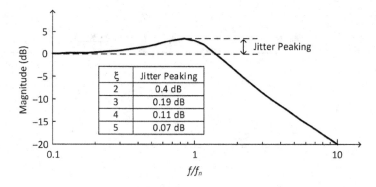

Figure 7.7 Jitter transfer function for three different values of ξ.

The jitter transfer function $H_T(s)$, as shown in Equation 7.6 has one zero s_z and two poles, s_{p1} and s_{p2}, all on the left-half plane:

$$s_z = -1/\tau$$

$$s_{p1} = \frac{-K\tau}{2} + \sqrt{(\frac{K\tau}{2})^2 - K} \qquad (7.11)$$

$$s_{p2} = \frac{-K\tau}{2} - \sqrt{(\frac{K\tau}{2})^2 - K}.$$

A sketch of this pole-zero location is shown in Figure 7.8 for different values of K. The zero always lies on the negative real axis. The poles, also, lie on the left-half plane but may or may not lie on the real axis, depending on whether K is above or below $4/\tau^2$, respectively. By design, we choose K so as to restrict the poles on the real axis, reducing the jitter peaking. However, as depicted in Figure 7.8, even in this case, the zero lies closer to the $j\omega$ axis than either of the two poles. This zero will cause the magnitude of the jitter transfer function to rise before it begins to fall, and this unavoidably creates peaking. As we further increase K, s_{p1} gets closer to s_z, effectively canceling s_z, leaving $H_T(s)$ with only one pole, s_{p2}, which is much farther away from the $j\omega$ axis. To see this, we will derive approximate expressions for s_{p1} and s_{p2} when $K >> 4/\tau^2$. Using Equation 7.11, we can write:

$$\begin{aligned} s_{p1} &= \tfrac{-K\tau}{2} + \tfrac{K\tau}{2}\left(1 - \tfrac{4}{K\tau^2}\right)^{1/2} \\ &\approx \tfrac{-K\tau}{2} + \tfrac{K\tau}{2}\left(1 - \tfrac{2}{K\tau^2}\right) \\ &= -1/\tau = s_z \end{aligned} \qquad (7.12)$$

$$s_{p2} \approx -K\tau.$$

Due to the cancellation (or near cancellation) of the first pole by the zero, the second pole, s_{p2} defines the bandwidth of the jitter transfer function. This can be observed in Figure 7.6, in which the jitter transfer bandwidth increases with increasing ξ or, equivalently, K.

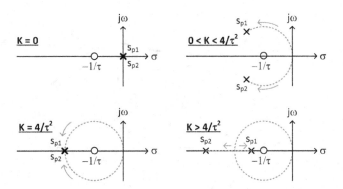

Figure 7.8 Pole-zero locations of the jitter transfer function for two values of ξ.

It has been shown [85] that the peaking can be eliminated if we are able to move the zero to the feedback path. Since the loop gain will remain the same, the loop stability will not be affected by this change; however, the jitter transfer function will no longer have a zero, which will eliminate peaking altogether.

7.2.3   Jitter Generation

If the input data to the CDR is clean with no jitter, i.e., $\varphi_{IN} = 0$, the jitter of the recovered clock comes directly from the VCO jitter. The transfer function that relates the VCO jitter to the recovered clock jitter is known as jitter generation. For the CDR model shown in Figure 7.5, and using the notation introduced earlier, this transfer function can be written as:

$$H_G(s) = \left.\frac{\varphi_{out}}{\varphi_{VCO}}\right|_{\varphi_{in}=0} = \frac{s^2}{s^2 + K\tau s + K} \tag{7.13}$$

or, in canonical form, using the notation introduced in Equation 7.7:

$$H_G(s) = \left.\frac{\varphi_{out}}{\varphi_{VCO}}\right|_{\varphi_{in}=0} = \frac{s^2}{s^2 + 2\xi\omega_n s + \omega_n^2}. \tag{7.14}$$

Replacing s with $j\omega$ ($= j2\pi f$), we can derive the jitter generation function as follows:

$$|H_G(f)|^2 = \frac{\omega^4}{(\omega_n^2 - \omega^2)^2 + 4\xi^2\omega_n^2\omega^2}. \tag{7.15}$$

Referring to Equation 7.13, jitter generation is a high-pass filter with two zeros, at zero frequency, and two poles identical to those of the jitter transfer function (see Equation 7.11). A Bode plot of jitter generation, as shown in Figure 7.9, clearly reveals two changes of slope (from +40dB/dec to +20dB/dec, and then to 0dB/dec) in the jitter generation magnitude, corresponding to its two poles. In contrast, we do not observe a 40dB/dec slope in the jitter transfer function simply because its first pole is cancelled (or almost cancelled) by its zero and it is considered a single-pole system. Nevertheless, we observe that the corner frequency (or the 3dB frequency) of both functions correspond to the same second pole. A smaller second pole will reduce jitter transfer, which

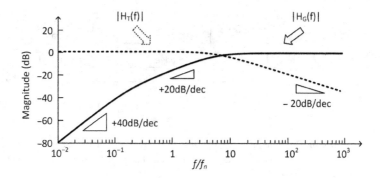

Figure 7.9 Jitter transfer and jitter generation curves.

is desirable, but at the same time, it increases jitter generation, which is not desirable. To decouple the two transfer functions, we can resort to the CDR architecture discussed in the previous section where the zero in the feed-forward path is moved to the feedback path [85].

7.2.4   Jitter Tolerance

An important figure of merit for a CDR is its tolerance to input jitter, that is, the maximum allowable amplitude of input jitter that keeps the bit error rate (BER) of the CDR below a target value, say 10^{-12} or 10^{-15}. To quantify jitter tolerance, we often apply a sinusoidal jitter of a fixed frequency to the CDR input data and observe the BER of the CDR. As we increase the amplitude of this sinusoidal jitter while keeping its frequency constant, there will be an amplitude at which the BER increases beyond an acceptable value. We refer to this particular jitter amplitude the jitter tolerance at the corresponding jitter frequency. As we repeat this experiment for different jitter frequencies, we can plot jitter tolerance as a function of jitter frequency. This plot is known as jitter tolerance.

In the past few sections, we have emphasized that jitter is a random signal and needs to be treated as such. However, in the definition of jitter tolerance, we are applying "sinusoidal" jitter. Clearly, this is not a random signal, and hence the jitter tolerance curve does not capture a CDR's true tolerance to random jitter. We note, however, that we can treat sinusoidal jitter as a deterministic signal, so that we do not need to evoke the PSD of the jitter, but instead deal only with the jitter's amplitude and frequency. Given this, let us now derive a formula for jitter tolerance, which we will represent as $JTOL(f)$. Referring to Figure 7.4, we can write:

$$JTOL(f) = |\varphi_{in}(f)|_{pp-max} \quad \text{for a fixed } BER \tag{7.16}$$

where the subscript $pp-max$ indicates the maximum peak-to-peak amplitude. We can further expand this equation as follows:

$$JTOL(f) = \left| \frac{\varphi_{in}(f)}{\varphi_e(f)} \right| \cdot |\varphi_e(f)|_{pp-max} \tag{7.17}$$

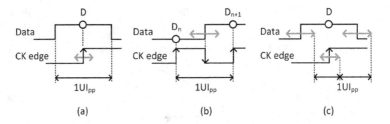

Figure 7.10 Relative jitter must be less than 1UIpp for error-free operation under all circumstances, (a) ideal data with jittery clock, (b) jittery data with ideal clock, (c) jittery data and jittery clock.

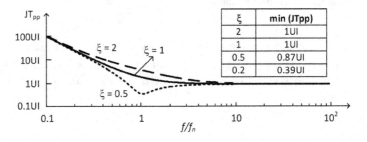

Figure 7.11 An ideal jitter tolerance curve for a second-order CDR.

In an ideal CDR, the maximum peak-to-peak amplitude of $|\varphi_e(f)|$ is 1UI. This can be seen with the aid of Figure 7.10, where in (a), the relative jitter is applied to the clock edge while the data edge is kept jitter free, and in (b), the situation is reversed. In (a), the clock edge may only move by 1UIpp around its nominal location or else a bit error would occur. Similarly, in (b), the data edge may only move by 1UIpp around its nominal location or else wrong data is sampled by the clock edge.

Accordingly, jitter tolerance can be expressed in terms of the number of UIs as:

$$JTOL(f) = \left| \frac{\varphi_{in}(f)}{\varphi_e(f)} \right| \quad \text{[UI].} \tag{7.18}$$

Given the CDR loop shown in Figure 7.4, we can write:

$$JTOL(f) = \left| 1 + \frac{K_{PD}K_{VCO}H_{LF}(f)}{j2\pi f} \right| \quad \text{[UI]} \tag{7.19}$$

where the second term inside the brackets is the loop gain of the CDR. If we now expand $H_{LF}(f)$ for the CDR of Figure 7.5, we can write:

$$JTOL(f) = \left| 1 - 2\xi j \left(\frac{f_n}{f} \right) - \left(\frac{f_n}{f} \right)^2 \right| \quad \text{[UI]} \tag{7.20}$$

A plot of the jitter tolerance as a function of jitter frequency is shown in a log–log scale for three different values of ξ in Figure 7.11.

At frequencies far below and above the natural frequency, the jitter tolerance can be approximated by the following:

$$JTOL(f) = \begin{cases} \left(\frac{f_n}{f}\right)^2 & \text{if } f << f_n \\ 1 & \text{if } f >> f_n. \end{cases} \tag{7.21}$$

Equation 7.21 states that the jitter tolerance at very high jitter frequencies is limited to 1UIpp. This is consistent with our observation earlier that the recovered clock does not track the high-frequency jitter, limiting the maximum peak-to-peak deviation of the data edge from its nominal position to 1UI. Any further deviation will result in the recovered clock missing a bit, increasing the BER. On the other end of the spectrum, for jitter frequencies below f_c, the jitter tolerance is increased at 40dB/decade (corresponding to the f^2 term) as we reduce the jitter frequency. Again, this is consistent with our observation earlier that the recovered clock better tracks data jitter at lower jitter frequencies or, equivalently, that the data edge and the clock edge move together in the same direction. As a result, the relative jitter between the data and the clock remains small (i.e., below 1UI peak-to-peak) even when the data jitter amplitude is much larger than 1UI. Around the corner frequency, the jitter tolerance dips below the 1UIpp for underdamped systems, i.e., when $\xi < 1$. Figure 7.11 shows the minimum jitter tolerance (in UIpp) for four distinct values of ξ.

The jitter tolerance as shown in Figure 7.11 is under ideal conditions. For example, it assumes that the input data has no jitter other than the sinusoidal jitter, that there are zero jitter contributions from the VCO, and that the circuit elements in the CDR are ideal. In reality, none of these conditions is met: the input data often contains some residual ISI jitter, i.e., the jitter that remains after the receiver equalization, in addition to some crosstalk-induced jitter due to the interference with data from adjacent channels, and some random jitter from the transmitter clock. The VCO will undoubtedly generate further random jitter, and finally there will be noise from various elements in the circuits, such as from the RC filter and from the power supply. These non-idealities will reduce the jitter tolerance at all frequencies, but their effect is more observable at high frequencies where the jitter tolerance is below 1UIpp. A typical value for the high-frequency jitter tolerance can be any number between 0.2UIpp to 0.7UIpp. A design that can tolerate a higher jitter may have been over-designed, e.g., consuming too much power, and a design whose jitter tolerance is below 0.2UIpp may be too risky, i.e., does not have enough jitter margin to be used in a product.

Effect of Limited VCO Tuning Range

In the derivation of the equations for the jitter tolerance, we have assumed the VCO to have an unlimited tuning range. In other words, we have characterized the VCO by $f_{VCO} = f_{nom} + K_{VCO}V_{cntl}$, where V_{cntl} is the control voltage of the VCO in Figure 7.5 and f_{nom} is the nominal VCO frequency corresponding to $V_{cntl} = 0$. We now consider the case where the VCO frequency is limited to $\pm\Delta f_p$ around f_{nom}, i.e.,

$$f_{nom} - \Delta f_p < f_{VCO} < f_{nom} + \Delta f_p. \tag{7.22}$$

If we represent the input phase with φ_{in} and the sinusoidal jitter as $A_m sin(\omega_m t)$, then we can write:

$$\varphi_{in}(t) = 2\pi f_{nom}t + A_m sin(\omega_m t). \qquad (7.23)$$

Taking the derivative of both sides provides the instantaneous input frequency, f_{in}:

$$f_{in}(t) = \frac{1}{2\pi}\frac{d}{dt}\varphi_{in}(t) = f_{nom} + A_m f_m cos(\omega_m t). \qquad (7.24)$$

And, therefore, the range of f_{in} is limited to:

$$f_{nom} - A_m f_m < f_{in} < f_{nom} + A_m f_m. \qquad (7.25)$$

Comparing Equations 7.25 and 7.22, a necessary condition for CDR locking is $A_m f_m \leq \Delta f_p$, where A_m is expressed in radians. Therefore, the peak-to-peak jitter amplitude, which would be $2A_m$ in radians, can be written in UIs as follows:

$$JTOL(f) \leq \frac{\Delta f_p}{\pi f} \qquad (7.26)$$

where we have replaced f_m with f to be consistent with the notation used earlier in $JTOL(f)$. According to Equation 7.21, for a VCO with an unlimited tuning range, the jitter tolerance increases by 40dB/decade as the jitter frequency decreases. This equation, however, states that if the VCO has a limited tuning range, the jitter tolerance can increase only at 20dB/decade as jitter frequency decreases. If we superimpose these two functions, as shown in Figure 7.12, we can see that Equation 7.26 provides a lower limit to jitter tolerance.

This limit is due only to the limited tuning range of the VCO and can increase with the tuning range.

Jitter Tolerance Measurement

A basic setup for jitter tolerance measurement is shown in Figure 7.13. A sinusoidal signal whose amplitude and frequency can be adjusted modulates the frequency of the clock waveform, effectively adding jitter with adjustable amplitude and frequency to the

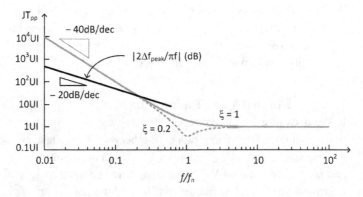

Figure 7.12 Jitter tolerance is limited at low jitter frequencies by the limited tuning range of the VCO.

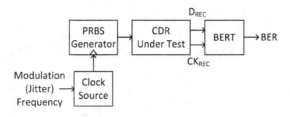

Figure 7.13 A basic measurement setup for jitter tolerance.

clock edges. This jittery clock is then applied to a pseudo random bit sequence (PRBS) generator to produce jittery data and test the CDR. To plot jitter tolerance as a function of jitter frequency, we increase the sinusoidal jitter amplitude in small steps until the bit-error-rate tester (BERT) indicates a BER higher than the target BER, say 10^{-12}. We then record the highest jitter amplitude that meets the BER target as the jitter tolerance at that particular jitter frequency. Obviously, we need to repeat this measurement for various jitter frequencies in order to plot the function. Since jitter frequency can span several decades, this measurement is performed for a few jitter frequencies (say 5–6) per frequency decade.

7.3 Effect of Jitter on Bang-Bang CDR

In the previous section, we explained how jitter is shaped as it goes through a CDR. In this section, we show how jitter can affect the system dynamics of a bang-bang CDR and, in particular, how jitter can be used to improve system stability.

7.3.1 Background

A bang-bang CDR is similar to a linear CDR as shown earlier in Figure 7.3 except for the implementation of its phase detector, which is called a *bang-bang phase detector* (BB-PD). A typical implementation of a BB-PD, followed by a charge pump and an RC loop filter, is shown in Figure 7.14(a). The bang-bang phase detector samples the incoming data twice per UI, as shown in Figure 7.14(b), once with the rising edge and once with the falling edge of the recovered clock (assuming a full-rate clock, i.e., a clock with the period equal to 1UI), producing two bits in every clock cycle. The bit corresponding to the rising edge of the clock is called the data bit (or the center bit) and the bit corresponding to the falling edge of the clock is called the edge bit (or the boundary bit). By comparing the edge bit against its two adjacent data bits, one can conclude if the recovered clock is early, corresponding to the edge bit matching the current data bit, or late, corresponding to the edge bit matching the next data bit. When the clock is early, the PD asserts its Early output, causing the charge pump to pull charge from the control node of the VCO, effectively reducing the VCO frequency. When the clock is late, the PD asserts its Late output, causing the charge pump to push charge to the control node of the VCO, effectively increasing the VCO frequency. Given the

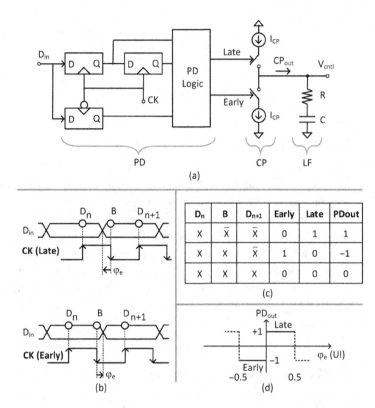

Figure 7.14 (a) A bang-bang PD followed by a charge pump and an RC loop filter, (b) bang-bang PD operation, (c) bang-bang PD logic, (d) bang-bang PD characteristics.

feedback loop, the VCO will adjust its phase over time so that the falling edge of the recovered clock is eventually aligned with the data edge, and thus the rising edge is the center of the data bit. This is when the CDR achieves *phase lock*.

BB-PD operation is summarized in the table shown in Figure 7.14(c). Note that the PD output, defined as Late minus Early, assumes 1 or -1 when the clock is late or early, respectively. When there is no transition in the input data, corresponding to the last row of the table, none of the Early or Late signals will be asserted. In this case, the PD output will be zero, leaving the VCO control voltage and frequency intact. In summary, the PD output can be defined as follows:

$$PD_{out} = \text{sgn}(\varphi_e) = \begin{cases} -1 & \text{if } \varphi_e < 0 \\ 1 & \text{if } \varphi_e > 0 \\ 0 & \text{if } \varphi_e = 0 \text{ (no data transition)} \end{cases} \quad (7.27)$$

where φ_e represents the phase error and is defined as the data phase minus the clock phase if there is data transition, and is assigned 0 otherwise.

If we assume the data transition density is α_T, then the expected value of the PD_{out} for a late or early clock will be $+\alpha_T$ or $-\alpha_T$, respectively. In other words, we can write:

$$\text{E}[PD_{out}] = \alpha_T \text{sgn}(\varphi_e). \quad (7.28)$$

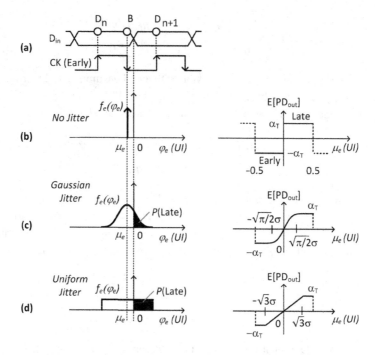

Figure 7.15 BB-PD characteristics under various jitter PDF. (a) BB-PD data and clock waveforms, (b) no jitter in phase error, (c) Gaussian jitter in phase error, and (d) uniform jitter in phase error.

An input–output characteristic of a BB-PD is shown in Figure 7.14(d). Clearly, the BB-PD has a nonlinear characteristic, with an infinite gain at the locking phase, and, as such, cannot use the linear model introduced in Figure 7.4. However, as we will show in this section, the relative jitter between the input data and the recovered clock will linearize the BB-PD characteristics around the locking phase, and this will allow us to use a linear model to gain insight into bang-bang CDR operation.

7.3.2   Effect of Jitter on BB-PD Gain

Figure 7.15 illustrates how the BB-PD gain varies as a function of both the jitter PDF and its standard deviation. Figure 7.15(a) repeats the nominal data and clock waveforms for when the clock is early. If there is no jitter in φ_e, the clock will always be early by the same amount, corresponding to a φ_e characterized by a delta function centered at a constant μ_e. In this case, the BB-PD will have an infinite gain around $\varphi_e = 0$. On the other hand, if we assume a nonzero jitter, say in the clock edge, then a nominally early clock may actually arrive late and vice versa. Figure 7.15(c) and (d) show this scenario for a φ_e with Gaussian and uniform PDF, respectively. In both cases, the expected value of φ_e is denoted by μ_e. φ_e is mostly negative, as its PDF is shifted to the left by μ_e, but there is a probability that φ_e can become positive, corresponding to the clock being late. This probability is equal to the area of the shaded region and its value depends on μ_e and the σ_e of the PDF. In other words:

$$P(PD_{out} = 1) \quad = \int_0^\infty f_e(\varphi_e)d\varphi_e = F(-\mu_e)$$
$$P(PD_{out} = -1) \quad = \int_{-\infty}^0 f_e(\varphi_e)d\varphi_e = 1 - F(-\mu_e)$$

(7.29)

where $f_e(\cdot)$ and $F_e(\cdot)$ represent the PDF and the Cumulative Distribution Function (CDF) of φ_e, respectively, and $F(\cdot)$ represents the CDF of the same distribution when centered around zero. The expected PD output can be calculated as:

$$E[PD_{out}] \quad = \alpha_T P(PD_{out} = 1) - \alpha_T P(PD_{out} = -1)$$
$$= \alpha_T(2F(-\mu_e) - 1).$$

(7.30)

This equation represents the BB-PD characteristics under an arbitrary jitter distribution. As depicted in Figures 7.15(c) and (d), the characteristics no longer have the sharp edge in the vicinity of the locking phase, but instead there is a finite slope which is a function of the jitter PDF alone. This slope can be easily derived as follows:

$$K_{PD} \quad = \frac{d}{d\mu_e}E[PD_{out}]$$
$$= 2\alpha_T f(-\mu_e).$$

(7.31)

The PD gain at locking point (i.e., when $\mu_e = 0$), K_{PD0}, can be written as:

$$K_{PD0} = \begin{cases} \sqrt{\dfrac{2}{\pi}}\dfrac{\alpha_T}{\sigma_e} & \text{for Gaussian distribution} \\[3mm] \sqrt{\dfrac{1}{3}}\dfrac{\alpha_T}{\sigma_e} & \text{for uniform distribution.} \end{cases}$$

(7.32)

Equation 7.32 clearly shows that jitter properties such as the PDF and the standard deviation determine the PD gain and, as we will see later in this section, will affect the dynamic behavior of the BB-CDR.

7.3.3   Added Jitter due to BB-PD

The BB-PD and its linear model are equivalent in the sense that, for the same input jitter φ_e, their outputs will have the same expected values. However, they are not equivalent in terms of their output jitter variances. To correct for this, we need to put back additional jitter φ_{PD} into our model as shown in Figure 7.16. This jitter plays the same role as that of the quantization error in an ADC linearization process. Given $\varphi_{PD} = \text{sgn}(\varphi_e) - K_{PD0}\varphi_e$, let us now find the expected value and the standard variation of φ_{PD} with Gaussian distribution.

$$E[\varphi_{PD}] = E\left[\text{sgn}(\varphi_e)\right] - E[K_{PD0}\varphi_e] = 0.$$

(7.33)

This is in fact consistent with our definition of K_{PD0} as derived in Equation 7.32.

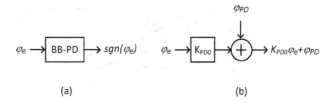

Figure 7.16 (a) Block diagram of a BB-PD, (b) linear equivalent model of a BB-PD.

$$
\begin{aligned}
\sigma_{PD}^2 &:= E\left[\varphi_{PD}^2\right] \\
&= E\left[\text{sgn}^2(\varphi_e)\right] + E\left[K_{PD0}^2 \varphi_e^2\right] - 2K_{PD0}E\left[\varphi_e \text{sgn}(\varphi_e)\right] \\
&= \alpha_T + K_{PD0}^2 \sigma_e^2 - 2K_{PD0}E\left[\varphi_e \text{sgn}(\varphi_e)\right] \\
&= \alpha_T - K_{PD0}^2 \sigma_e^2 \\
&= \alpha_T - \frac{2}{\pi}\alpha_T^2
\end{aligned}
\tag{7.34}
$$

where we have used Equation 7.32 corresponding to Gaussian distribution, and $E\left[\varphi_e \text{sgn}(\varphi_e)\right] = \sqrt{2/\pi}\alpha_T \sigma_e$, to derive the final expression for σ_{PD}^2.

An alternative method to find K_{PD0} and σ_{PD} is to begin from $\varphi_{PD} = \text{sgn}(\varphi_e) - K_{PD0}\varphi_e$ and try to minimize $E\left[\varphi_{PD}^2\right]$. This method as described in [86], [87] results in the same set of equations as we derived here.

7.3.4 Effect of Jitter on BB-CDR Stability

Our linear model of the CDR (Figure 7.5) neglected the effect of the VCO's input capacitance in derivations of various transfer functions. As this parasitic capacitance introduces a new pole into the system, it plays an important role in CDR stability. We now derive an expression for the loop gain of the CDR in order to study its stability.

Let us denote by C_p the parasitic capacitance at the VCO input. This capacitor appears in parallel with the series RC circuit of the loop filter as shown in Figure 7.5. Taking C_p into account, we can write an equation for the loop gain as follows:

$$
LG(s) = K\frac{1 + \tau s}{s^2(1 + \tau_p s)}
\tag{7.35}
$$

where K is now defined as $K_{PD}K_{VCO}/(C_p + C)$ and $\tau_p = RCC_p(C + C_p)$. The loop gain has a zero at $-1/\tau$, two poles at zero, and one nonzero pole due to the VCO parasitic capacitance at $-1/\tau_p$. We will show that, depending on K, which is directly influenced by the jitter PDF and its standard deviation σ (as shown in Equation 7.32), the system may lose or gain phase margin and, as such, move away from or towards stability.

Let us denote the unity-gain frequency of the loop gain with ω_u. Then we can write an expression for the phase margin PM of this system as follows:

$$
PM = \tan^{-1}(\tau\omega_u) - \tan^{-1}(\tau_p\omega_u).
\tag{7.36}
$$

Note that τ and τ_p in this equation are set by design. ω_u, however, is directly influenced by K, which in turn is influenced by the jitter PDF and its σ. To illustrate this, Figure 7.17 shows the Bode plots of three loop gains, each for a different jitter PDF and σ.

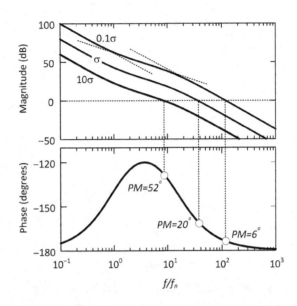

Figure 7.17 Bode plot of the CDR loop gain illustrating an increase in phase margin RMS jitter increases.

7.4   Jitter in the Received Eye

A taxonomy of jitter was presented earlier in Section 2.2.6. In this section, we illustrate the jitter components in wireline applications as they manifest themselves in eye diagrams. Figure 7.18 shows four different eye diagrams associated with the same link but under different jitter conditions. To begin, Figure 7.18(a) shows an ideal eye diagram with almost no jitter. The corresponding jitter histogram shows a single peak that corresponds to the expected zero crossing location.

Figure 7.18(b) shows an eye diagram with Gaussian random jitter where jitter is "spread" in time and disappears gradually as we move away from the nominal zero-crossing position. This jitter is unbounded and attributed mainly to thermal noise and flicker noise in the circuits, but can also be attributed to the combined characteristics of several uncorrelated noise sources.

In contrast with unbounded jitter, we have bounded jitter, which includes period jitter, sinusoidal jitter, data-dependent jitter, and ISI jitter. As an example of bounded jitter, Figure 7.18(c) shows the histogram of sinusoidal jitter and its impact on the eye diagram. The sinusoidal jitter could be due to power-supply noise or it may have been injected intentionally for the purpose of measuring of jitter tolerance.

ISI will cause the zero crossings to move as a function of the bit sequence and limited channel bandwidth, resulting in several zero crossings, creating discrete peaks. Figure 7.18(d) shows an example of an eye under the influence of ISI only.

Jitter can also be caused by the clock's duty cycle distortion in double-data-rate (DDR) systems. Figures 7.19(a) and 7.19(b) show two versions of the same eye diagram at different zoom levels, with (a) showing the jitter distribution at zero crossings

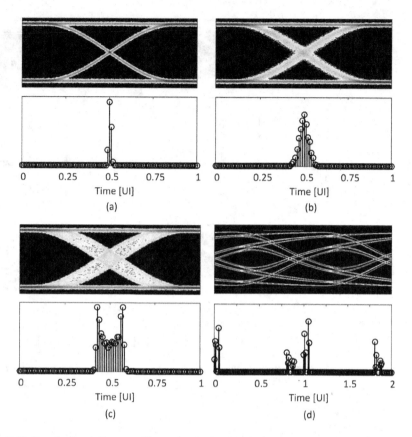

Figure 7.18 Received eye diagrams illustrating various jitter components, (a) no jitter, (b) Gaussian jitter, (c) sinusoidal jitter, (d) ISI-induced jitter.

over one UI and (b) showing the corresponding distributions of several zero crossings over four UIs. Different rise and fall times, or a receiver voltage threshold that is not centered, will create a jitter similar to that of a DCD, as shown in Figure 7.19(c). Note that, in this figure, the voltage threshold level of the receiver is assumed to be centered, creating a distribution of zero-crossings on the two sides of the crossing shown in the eye. A similar eye would occur if the rise and fall times were the same, but the voltage threshold were not centered.

In real systems, two or more noise sources may affect the eye diagram. Figure 7.19(d) shows the effects of random jitter and DCD at the same time. The resulting jitter distribution is the convolution of the distributions of the two jitter sources. For further review, refer to Section 2.2.7.

7.5   Jitter Amplification by Passive Channels

As mentioned earlier, a typical channel in a wireline system attenuates the high-frequency components of the transmit signal. However, if the channel is used to

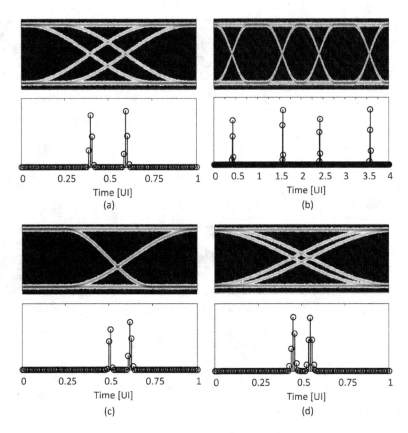

Figure 7.19 Received eye diagrams illustrating DCD combined with random jitter, (a) DCD-induced jitter, (b) DCD-induced jitter over four UIs, (c) jitter due to rise/fall time asymmetry, (d) jitter due to the combined effect of DCD and random jitter.

transmit a clock, such as in a clock-forwarding architecture, the clock jitter at the end of the channel (received clock) is typically higher than the jitter in the transmit clock. We refer to the ratio $\sigma_{RCK}/\sigma_{TCK}$ as the jitter amplification factor where σ_{RCK} and σ_{TCK} refer to the standard deviation of the jitter in the received clock and the transmit clock, respectively.

Jitter amplification may seem counterintuitive at first because passive channels are not expected to amplify any signal, as they are simply made up of wires with no active elements. This becomes more intuitive, however, if we distinguish between the signal transfer function of a channel and its jitter transfer function. While the signal (typically voltage) transfer function of a wireline channel is a low-pass filter and never amplifies any signal, at any frequency, the jitter transfer function of the same channel becomes a high-pass filter and could amplify jitter. While several papers have discussed the topics of jitter amplification [88], [89], [90], [91], both in the time domain and frequency domain, and included measured results, simulations, and theoretical analyses, we will mainly follow the procedure in [88] to explain this concept in the frequency domain.

Consider a simplified waveform for the transmit clock contaminated by a sinusoidal jitter (or excess phase):

$$v_{TCK}(t) = A \cos\left(2\pi f_c t + A_j \sin(2\pi f_j t)\right) \tag{7.37}$$

where f_c is the clock frequency, A_j is the excess phase amplitude (or jitter amplitude in radians), and f_j is the jitter frequency. If we assume the jitter amplitude is much smaller than 2π, then we can approximate this equation by the following:

$$v_{TCK}(t) = A \cos(2\pi f_c t) - \frac{A A_j}{2}\left(\cos(2\pi f_L t) - \cos(2\pi f_H t)\right) \tag{7.38}$$

where $f_H = f_c + f_j$ and $f_L = f_c - f_j$ [88]. Taking the Fourier transform of both sides of this equation, we obtain an expression for the transmit clock in the frequency domain:

$$V_{TCK}(f) = \frac{A}{2}\delta(f - f_c) - \frac{A A_j}{4}\left(\delta(f - f_L) - \delta(f - f_H)\right) \tag{7.39}$$

where we have shown only the positive frequency portion of the spectrum; the spectrum is symmetrical with respect to frequency. We are now ready to subject this expression of the transmit clock to the voltage transfer function of the channel, which we will denote by $H(f)$, in order to find an expression for the received clock:

$$V_{RCK}(f) = \frac{A}{2}H(f_c)\delta(f - f_c) - \frac{A A_j}{4}\left(H(f_L)\delta(f - f_L) - H(f_H)\delta(f - f_H)\right). \tag{7.40}$$

In doing so, we have assumed the channel is a linear-phase channel so as to neglect additional jitter that may be caused by a nonlinear phase response. Given that $H(f_c), H(f_L)$, and $H(f_H)$ are all constants, we can simply take the inverse Fourier transform of this equation and, after similar approximations to those made for the transmit clock, arrive at a time-domain expression for the received clock:

$$v_{RCK}(t) = H(f_c)A\cos\left(2\pi f_c t + \frac{H(f_L) + H(f_H)}{2H(f_c)}A_j \sin(2\pi f_j t)\right). \tag{7.41}$$

This equation simply states that the channel attenuates the clock amplitude by a factor of $H(f_c)$ while multiplying its sinusoidal jitter by a factor of $[H(f_L) + H(f_H)]/[2H(f_c)]$. If the channel is relatively flat around f_c, i.e., if $H(f_L) \approx H(f_H) \approx H(f_c)$, then the jitter amplification factor is around 1, and hence the jitter in the received clock will be about the same jitter as in the transmit clock. However, if we operate the channel at higher frequencies, the channel exhibits steep frequency roll off (such as an exponential roll off $e^{-\alpha f}$ around f_c), we will have $H(f_L) \gg H(f_c) \gg H(f_H)$, and this results in a jitter amplification factor that is higher than 1 (on the order of 2 to 5).

7.6   Jitter Monitoring and Mitigation

In a wireline link, the transmit data accumulates jitter of various types until it arrives at the receiver, where it is fed to the equalizer and subsequently to the CDR. The jitter that matters the most at the end is φ_e (shown in Figure 7.4), which is the relative jitter at the input of the phase detector of the CDR, that is, the absolute jitter of the equalized

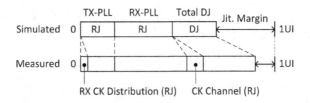

Figure 7.20 An example jitter budget [92] simulated at design time versus the actual jitter budget measured after fabrication.

data, φ_{in} minus the absolute jitter of the recovered clock, φ_{out}. It is this relative jitter that impacts the BER of the receiver. The absolute jitter of the equalized data includes the TX PLL jitter, the channel-induced jitter as described in Section 7.5, and the data-dependent jitter, which is partially removed by the equalization at the receiver. The recovered clock includes the RX PLL jitter, the VCO jitter of the CDR, as well as the equalized data jitter shaped by the CDR loop. The peak-to-peak relative jitter should never exceed 1UI or there will be no position in the UI left at which to sample this jittery data. In fact, we should reserve some interval within the 1UI period for jitter tolerance, just in case some of the components contribute more jitter than expected. To illustrate this, we show in Figure 7.20, a simplified example of a jitter budget for a 25Gb/s optical link [92] and compare various jitter contributions at the time of design with the actual jitter contributions as collected during measurement.

This example shows how simulations underestimated some jitters, such as the jitter from the RX PLL and the ISI-induced jitter, overestimated the TX PLL jitter, and totally neglected the clock distribution jitter and the jitter from the clock channel. While the design was expected to have a jitter margin (i.e., high-frequency jitter tolerance) of 0.3UI, the measured jitter margin was only about 0.1UI. It is in this context that jitter monitoring and measurement become important. Jitter monitoring and measurement can achieve two goals: 1. they enable us to better calibrate our circuits for future designs by incorporating more realistic jitter characteristics into our models and simulations (this, in turn, will lead to a better estimate of the jitter tolerance of the fabricated design), and 2. they will enable us to adaptively change certain parameters within the system so as to mitigate jitter and its effect on CDR performance by either reducing the jitter itself or reducing its effect on BER. In this section, we will examine several methods of jitter monitoring and mitigation, which are two active areas of research in wireline.

7.6.1   Eye-Opening Monitor

An Eye-Opening Monitor (EOM) is a simple circuit that allows monitoring of the data eye at the receiver input and producing information such as the eye opening and the jitter histogram. Figure 7.21 shows a block diagram of an EOM implemented for a CDR operating at 40Gb/s [93].

The EOM operation is best described if we first assume that the CDR is in a phase lock position, i.e., if we assume that the recovered clock samples the received data waveform at the center of the eye using the CDR sampler. In this case, we consider both the phase

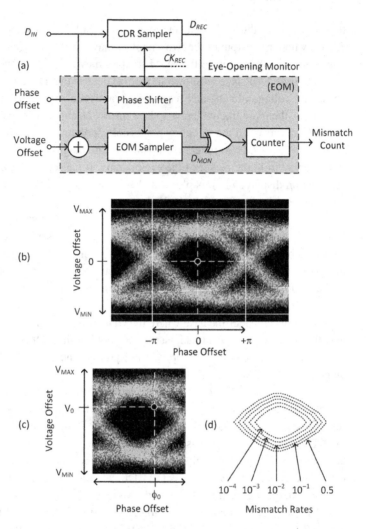

Figure 7.21 (a) A block diagram of an eye-opening monitor, (b) monitoring the eye with zero voltage and phase offsets, (c) monitoring the eye as we sweep the voltage and phase offsets, (d) eye contours corresponding to different BERs.

of the recovered clock and the voltage threshold of the CDR sampler to be at their ideal values. The EOM samples the received data waveform at a phase offset with respect to the recovered clock and voltage offset with respect to the threshold of the CDR sampler. The eye opening is then monitored by measuring the difference in the outputs of the two samplers. Since we have assumed the CDR is in an ideal lock position, the point corresponding to the phase of the recovered clock and the CDR sampler threshold corresponds to the center of the eye as shown in Figure 7.21(b). If the phase shift and the voltage offset in the EOM block are small enough, the outputs of the two samplers agree with each other, and that means the sampling points of both the CDR sampler and the EOM sampler are within the eye opening area of the eye diagram. As we increase either the phase shift or the voltage offset, however, we will reach a point, as depicted in Figure 7.21(c) where the output of the two samplers may not always agree, i.e., due to jitter and noise, the sampling point of the EOM sampler may fall outside the opening area of the eye. A simple XOR gate as shown in part (a) detects this mismatch as a function of the phase and voltage offsets, and feeds this to a mismatch counter. Obviously, the larger the offsets, the more likely there is to be a mismatch and, hence, the larger the mismatch count. Contours of all the sampling points corresponding to the same mismatch rates are shown in Figure 7.21(d) where the mismatch rate ranges from 10^{-4} to 0.5 as we move farther away from the center of the eye opening. If we set the offset voltage to zero, the information displayed in this plot can also be used to plot the BER as a function of the phase offset, also known as the bathtub curve.

The EOM as described above is also called a "two-dimensional" EOM to distinguish it from a "one-dimensional" EOM where the voltage offset is fixed, say to 0, and only the phase offset is swept. In this case, we can essentially measure the CDF (and hence the PDF) of the jitter for a fixed voltage offset. The silicon area cost, and the accuracy of the resulting PDF, depends on the resolution by which we are able to increment the phase shift.

Now, let us consider the case where the CDR is not in a perfect lock position and see how the EOM can be used to adjust the threshold of the CDR sampler and the phase of the recovered clock so as to improve CDR performance. Figure 7.22(a) shows a block diagram of a CDR along with the EOM.

Here, an off-chip algorithm implemented on a PC monitors the outputs of the CDR sampler and the EOM sampler and adjusts the voltage thresholds and the phase offsets for both samplers, not just for the EOM sampler. To see how this works, let us assume the CDR sampling position is off-center initially, as depicted in Figure 7.22(b). The algorithm then sweeps the voltage offset and the phase offset for the EOM sampler so as to find the eye opening with respect to CDR sampling position. Once the eye opening is determined, the algorithm will adjust the voltage offset and the phase shift for the CDR sampler such that its sampling position is moved to the center of the eye, as shown in Figure 7.22. This technique is claimed [93] to improve the BER from 10^{-7} to 10^{-12} at an input power of -21dBm.

7.6.2   Relative Jitter Measurement

Aside from the EOM presented in the previous section, relative jitter can be monitored by either a time-to-digital converter (TDC) as shown in Figure 7.23(a) or an ADC at the output of a linear PD as shown in Figure 7.23(b). In the former approach, the TDC could be an extra block in addition to a PD already present in the CDR. The latter approach is one way of building a TDC by feeding the output of an existing PD to an ADC. However, for the ADC output to linearly represent the relative jitter, the PD must be linear; otherwise, the TDC output will be a distorted (nonlinear) representation of the relative jitter. [94] and [95] present a good treatment of these cases, but we will focus here on monitoring the relative jitter in a BB-CDR in order to improve its jitter tolerance.

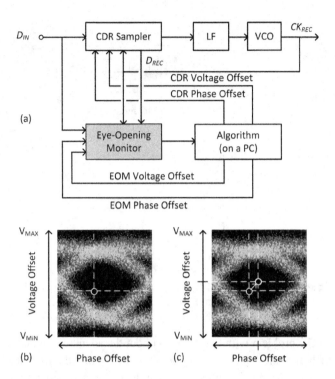

Figure 7.22 (a) Using EOM to find the best sampling position for the CDR, (b) initial sampling position of the CDR is not centered, (c) EOM is used to bring the sampling position to the center of the eye.

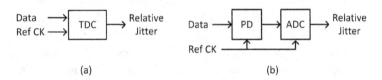

Figure 7.23 Relative jitter measurement (a) using a time-to-digital converter (b) using a linear PD followed by an ADC.

Figure 7.24 shows a simplified block diagram of a BB-CDR employing jitter monitoring to adaptively change the loop gain (K) for optimal jitter tolerance [96]. Without the adaptation block, the BB-CDR operates as follows. A BB-PD, as descried earlier in this chapter, counts the early/late events, and feeds the count to a gain block (K) and subsequently to a digital loop filter (LF), which contains both a proportional path and an integrative path similar to a charge pump and an analog loop filter. The LF then produces a phase code which will be used by the phase interpolator (PI) to adjust the phase of a reference clock accordingly and to produce the recovered clock.

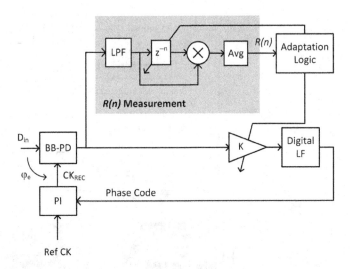

Figure 7.24 Monitoring jitter at the BB-PD output for optimizing the loop gain K. © 2017 IEEE. Reprinted, with permission, from [96].

The problem with this CDR, without the adaptation block, is that its loop dynamics, including the loop gain, the loop bandwidth, and the loop stability, are heavily dependent on the relative jitter, which in turn is influenced by the loop gain. To illustrate this, Figure 7.25 [96] plots the PSD of the relative jitter for three values of the loop gain K.

When K is less than or greater than desired, the relative jitter has a larger variance compared to the case of K desired. It is noted in [96] that a corresponding behavior can be observed in the autocorrelation function, $R(n)$, of the BB-PD output once it is low-pass filtered. $R(n)$, as plotted in Figure 7.25(b), shows an under-damped and an over-damped behavior for K greater than and less than desired, respectively. Accordingly, as shown in Figure 7.24, to find the desired K for optimal jitter performance, we simply observe $R(n)$, starting from a large value of K and reducing it until its oscillation subsides.

7.6.3   Absolute Jitter Measurement

Measuring the absolute jitter present in either a data or a clock waveform requires access to an ideal clock, i.e., a clock with no jitter. We can measure the absolute jitter in a waveform by simply feeding it along with its associated ideal clock into a linear phase detector. The output of the phase detector is then proportional to the absolute jitter of the waveform. However, in the absence of an ideal clock, we can only resort to other non-ideal waveforms to possibly extract information about the absolute jitter in each waveform.

To illustrate this, assume we are interested in the absolute jitter of data waveform A, φ_A, while we also have access to clock waveforms B and C with their uncorrelated absolute jitters of φ_B and φ_C, respectively. While we cannot observe φ_A, φ_B, and φ_C

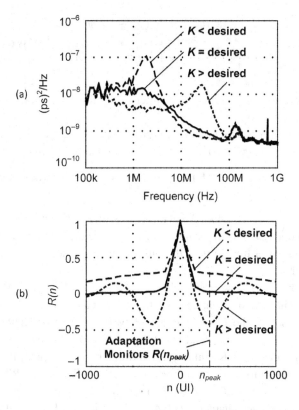

Figure 7.25 (a) Plots of PSD of relative jitter for different values of K: the desired K corresponds to minimum jitter variance, (b) plots of $R(n)$ for different values of loop gain K: the desired K corresponds to $R(n_{peak}) \approx 0$. © 2017 IEEE. Reprinted, with permission, from [96].

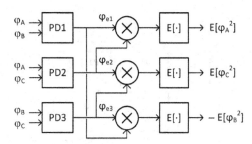

Figure 7.26 Basic block diagram of absolute jitter measurement. © 2015 IEEE. Reprinted, with permission, from [97].

directly, we can feed each pair of waveforms to a phase detector as shown in Figure 7.26 and observe their relative jitter at the corresponding PD output.

The PD outputs, φ_{e1}, φ_{e2}, and φ_{e3}, can then be correlated to produce the variances of the absolute jitters in the waveforms. For example, if we assume the PD is linear with a PD gain of one ($\varphi_{e1} = \varphi_A - \varphi_B$), we can write:

$$
\begin{aligned}
\mathrm{E}\left[\varphi_{e1}\varphi_{e2}\right] &= \mathrm{E}\left[(\varphi_A - \varphi_B)(\varphi_A - \varphi_C)\right] \\
&= \mathrm{E}\left[\varphi_A^2\right] + \mathrm{E}\left[\varphi_B\varphi_C\right] - \mathrm{E}\left[\varphi_A\varphi_C\right] - \mathrm{E}\left[\varphi_A\varphi_B\right] \\
&= \mathrm{E}\left[\varphi_A^2\right] \\
&= \sigma_A^2
\end{aligned}
\tag{7.42}
$$

where we have assumed the jitter in the three waveforms are all uncorrelated with a mean of zero.

For further insight into the absolute jitter, we can estimate its autocorrelation function by delaying φ_{e1} by n cycles prior to correlating it with φ_{e2}. Therefore, we can write:

$$
\begin{aligned}
\mathrm{E}\left[\varphi_{e1}(k-n)\varphi_{e2}(k)\right] &= \mathrm{E}\left[(\varphi_A(k-n)\varphi_A(k)\right] \\
&= R_{\varphi_A}(n)
\end{aligned}
\tag{7.43}
$$

where we have assumed that the jitter is wide-sense stationary so that the autocorrelation is a function only of the number of delay cycles, n, and not a function of the time index k. The Fourier transform of this autocorrelation function provides the PSD of the absolute jitter.

The scheme described above is implemented on chip [97] to estimate the absolute jitter in both the input data and the recovered clock in a 10Gb/s multi-lane CDR. Figure 7.27 shows the block diagram of two adjacent BB-CDR lanes employing identical BB-PDs. During normal operation, each CDR receives its own data and recovers the corresponding CK. During absolute jitter measurement, the two selectors at the front end are configured to feed Data1 to both CDR1 and CDR2 to produce CK1 and CK2. As such, CK1 and CK2 are both associated with the data waveform whose absolute jitter needs to be characterized. An additional PD, PD3, is added between the two lanes to further characterize the absolute jitter in each of CK1 and CK2.

To follow the procedure described earlier, the outputs of PD1, PD2, and PD3 need to be correlated in order to characterize the jitter in each of Data1, CK1, and CK2. Since

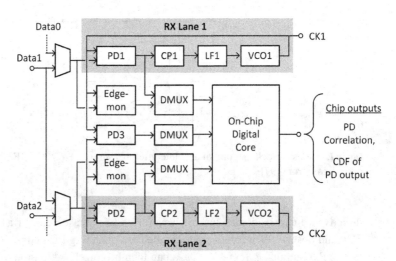

Figure 7.27 Absolute jitter measurement in two adjacent CDRs. © 2015 IEEE. Reprinted, with permission, from [97].

the PDs are bang-bang, their outputs are binary, and therefore the correlation can be easily performed on-chip in the digital domain using logic gates, FIFOs, and counters. The edge monitor blocks collect information on the jitter CDF so as to estimate the PD gain for each of PD1 and PD2. The data from the digital core is then sent off-chip to estimate the PSD of the absolute jitter in each of Data1, CK1, and CK2.

Let us now revisit our assumption about the correlation between Data1, CK1, and CK2. While the jitter in Data1 arises from jitter sources outside the CDR, the jitters in CK1 and CK2 are contributed by two sources: 1. the jitter generated by each of VCO1 and VCO2, and 2. the jitter in Data1. Since the two VCOs are assumed to be independent, or at least sufficiently separated so as to minimize any coupling between the two, and powered by well-isolated power supplies, the first contributions to CK1 and CK2 may indeed be considered uncorrelated. The second contribution comes from the same data jitter but it will be high-pass filtered by the loop before multiplying the PD outputs for the autocorrelation function. Therefore, the autocorrelation function for the absolute jitter in the data, as estimated by the block diagram in Figure 7.27, contains only the out-of-band portion of the jitter; the in-band portion of the jitter is heavily attenuated in the loop. Coincidentally, we are also interested more in the out-of-band jitter, as it is this portion of the jitter that degrades the CDR performance; the in-band jitter is tracked by the CDR loop.

Similarly, the absolute jitter in each of CK1 and CK2, as estimated by the above scheme, provides only a high-pass filtered version (the out-of-band portion) of the jitter in CK1 and CK2. One way to include more of the in-band jitter in this estimation is to reduce the loop bandwidth for diagnostic purposes [97].

7.7   Intentional Jitter

As mentioned in Section 7.3, the jitter PDF and its standard deviation affect the linearized gain of the BB-PD, which in turn impacts the loop dynamics of the CDR and its stability. In fact, it is possible to take advantage of this information and add intentional jitter to a CDR in order to gain linearity and other desired properties. In this section, we review two examples where we inject intentional jitter to improve PD linearity of a 10Gb/s CDR [98] and inject intentional jitter in a 28Gb/s CDR for jitter measurement [99].

7.7.1   Jitter Injection for Improved Linearity

Figure 7.28(a) shows the block diagram of a 10Gb/s CDR where the input data is sampled twice per UI, once at the data center using the data clock (CK_D) and once at the data edge using the edge clock (CK_E). To accommodate 10Gb/s operation, the clock is implemented by four phases of a quarter rate clock (2.5GHz). The data and edge samples are then demuxed by a factor of 8, feeding 32 data bits and 32 edge bits at the demuxed rate of 311MHz (10GHz/32) to a digital loop filter. By comparing the data and the edge bits, similar to what was described in Figure 7.14, the digital loop filter produces an

8-bit phase code for the data center and an 8-bit phase code for the data edge, with a nominal difference of 0.5UI between the two codes. Since these codes are used for a quarter-rate PI, the 8-bit phase code corresponds to a time resolution of 1.56ps. The jitter in the recovered clock in this CDR is influenced by two undesired effects. One is the finite resolution of the PI and the other is the bang-bang nature of the PD. Even though the phase code has a resolution of 8 bits, the DAC inside the PI is limited to 6 bits of resolution for area constraints. By intentionally adding a random 2-bit code to the phase code prior to dropping its 2 least significant bits, this work restores the effective resolution of the 6-bit code to 8 bits. In addition, for the edge clock only, a random code with a maximum of 0.5UIpp is added to its corresponding phase code to linearize the PD characteristics. As a result, the jitter added to the data clock will have a standard deviation that is different from the jitter added to the edge clock, as shown symbolically in Figure 7.28(b). This provides an extra degree of control for improving the PD linearity and the DAC resolution.

7.7.2   Jitter Injection for Jitter Measurement

We are interested in measuring the relative jitter between the data and the recovered clock in a bang-bang CDR by observing the bang-bang PD output. In Section 7.6.2, we used the autocorrelation function of the PD output and a linearized model of a BB-PD as shown in Figure 7.16, to arrive at the following expression for the relative jitter variance:

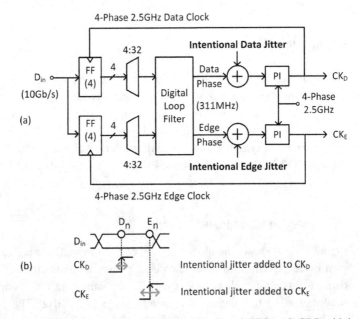

Figure 7.28 (a) A block diagram of a phase-interpolator-based (PI-based) CDR with intentional jitter in both the data clock and the edge clock, (b) data clock and edge clock waveforms with different jitter.

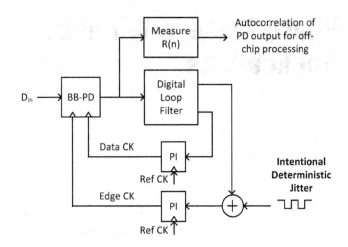

Figure 7.29 Adding intentional deterministic jitter to the edge clock enables measuring the relative jitter.

$$\sigma_e^2 = \sqrt{\frac{\alpha_T - \sigma_{PD}^2}{K_{PD0}^2}} \tag{7.44}$$

where α_T is the data transition density as defined in Section 7.6.2. Using this equation, we can estimate σ_e^2 provided we have the values for both σ_{PD}^2 and K_{PD0}. We can use Equation 7.34 to obtain σ_{PD}^2, and so it remains to evaluate K_{PD0} in order to determine σ_e^2. Evaluating K_{PD0}, however, is not trivial, because K_{PD0} is a function of σ_e itself, and therefore Equation 7.44 has two unknowns.

To add further observability to the PD output, [99] injects a small deterministic (square wave) jitter into the input of the phase interpolator controlling the edge clock jitter. This is shown in Figure 7.29. The injected jitter, which corresponds only to the least significant bit (LSB) of the PI control, is assumed to be much smaller than the relative jitter. As a result, the autocorrelation of the PD output will have two components: 1. the autocorrelation of the relative jitter, which is expected to be a delta function, corresponding to a mainly white relative jitter, and 2. the autocorrelation of the injected jitter, which is expected to be a triangular waveform. Given that the amplitude of the injected jitter is known and we can identify its contribution to the autocorrelation function of the PD output, we can determine K_{PD0} and hence estimate the power of the relative jitter using Equation 7.44. We refer the interested readers to [99] for further details of this scheme.

8 Phase Noise in Wireless Applications

A wireless transceiver typically uses a PLL at the transmitter to up-convert a baseband signal to an RF frequency and uses another PLL at the receiver to down-convert the RF signal back to the baseband. Depending on the quality of the PLL clock, which is often measured by its phase noise, the process of up-conversion and down-conversion produces unwanted signals which often interfere with signals in adjacent channels and deteriorate the SNR at the receiver. In this chapter, we first provide a brief background on the transmitter and receiver concepts, and then discuss the impact of PLL phase noise on transceiver performance.

8.1 Basics of Wireless Transceivers

Figure 8.1(a) shows the block diagram of a simplified wireless transmitter, in which a baseband signal is up-converted to an RF frequency in a two-step process: first to an intermediate frequency (IF), and then to an RF frequency, using a local oscillator (inside a PLL). The RF signal is then fed to a bandpass filter (BPF) followed by a power amplifier (PA) and the antenna. If the oscillator produces a pure single tone, the up-converted signal will simply contain the baseband signal translated (shifted) in frequency to f_{RF}. If the baseband signal has a bandwidth of B, the RF signal will have a bandwidth of 2B, as shown in Figure 8.1(b). However, due to the phase noise of the oscillator, the up-converted signal will also contain frequency content outside the 2B bandwidth, which may contribute to noise in adjacent receiver channels. The BPF allows the signal and the noise content within the 2B bandwidth to pass but attenuates the phase noise and other noise content that lie outside the 2B bandwidth. The PA amplifies the power of the RF signal so as to extend its reach to the intended receivers.

A simplified block diagram of a wireless receiver is shown in Figure 8.2(a). The received signal from the antenna is first filtered by an off-chip surface acoustic wave (SAW) filter, then passed to a low noise amplifier (LNA), and subsequently to a mixer that down-converts the RF signal either to an intermediate frequency (IF) or directly to DC for further filtering. The role of the BPF/LPF in the receiver is to select a channel among many channels that exist in the receiver band, similar to tuning into a radio station. Typical PSDs for the signals through the wireless receiver are shown in Figure 8.2(b).

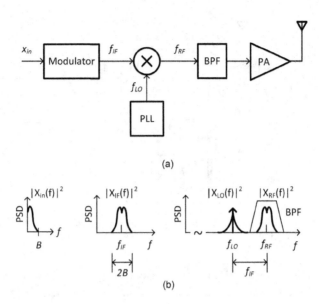

Figure 8.1 (a) A simplified block diagram of a wireless transmitter, (b) PSDs of the input, IF, LO, and RF signals.

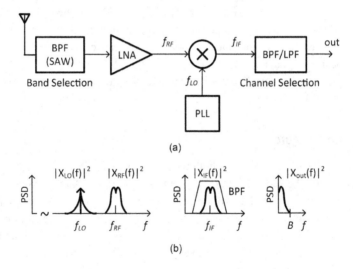

Figure 8.2 (a) A simplified block diagram of a wireless receiver, (b) PSD of the LO, RF, IF, and the output signals.

8.1.1 Blockers in Wireless Receivers

The desired received signal may occupy a bandwidth B of 100kHz (for example in GSM standard) and may be as weak as -102dBm (around $2.5\mu V_{rms}$). Undesired received signals, which lie outside the RX band and are referred to as out-of-band blockers, could be much higher in their power than the desired signal and, as such, they may prevent

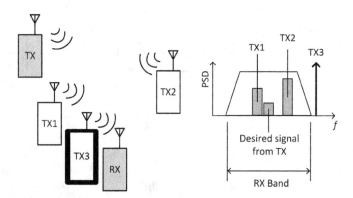

Figure 8.3 TX1 and TX2 are examples of in-band blockers, whereas TX3 is an example of an out-of-band blocker.

(block) the weak desired signal from being correctly detected. In addition to the out-of-band blockers, there are signals from adjacent channels in the same RX band, called in-band blockers, that can also create interference with the weak desired signal. The role of the SAW filter is to attenuate the out-of-band blockers and the role of filtering around IF is to attenuate in-band blockers. Nevertheless, a residual out-of-band blocker or in-band blocker may adversely affect the SNR at the receiver.

The out-of-band blockers, if not sufficiently attenuated, will force the receiver circuits to operate in a non-linear region simply due to their large amplitudes relative to the desired signals. Also, as we will see later, they can mix with other strong blockers and produce frequency components that directly overlap with the frequency spectrum of the desired signal.

8.1.2   Noise Figure

An ideal LNA amplifies both the desired signal and the noise, by the same factor, maintaining the same SNR at its output as at its input. A practical LNA, however, produces additional noise that will be added to its output. The added noise is generally produced by the active and passive components of the LNA. As a result, the SNR at the output of the LNA will be slightly lower than the SNR at the input of the LNA. The difference between the two SNRs, when expressed in dB, is defined as the noise figure (NF) for the LNA. In other words:

$$NF[\text{dB}] = SNR_{OUT}[\text{dB}] - SNR_{IN}[\text{dB}]. \tag{8.1}$$

A typical value for the NF of an LNA is around $2 - 3\text{dB}$.

8.1.3   Receiver Sensitivity

Receiver sensitivity is defined as the minimum power of the received signal at the receiver input that can be reliably detected in the presence of noise. As such, the receiver sensitivity is mainly constrained by the noise power at its input. Most receivers typically

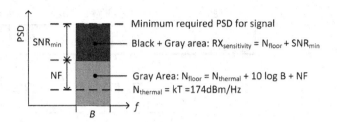

Figure 8.4 The RX sensitivity is defined as the minimum signal power reliably detectable by the receiver.

require a signal power level that is two to three times larger (or equivalently 3 to 5dB higher) than the noise power level. To determine the RX sensitivity, we typically determine the noise power at the input of the receiver and then multiply it by factor of 2 to 3 (i.e., add 3–5dB to it).

Looking at the block diagram of the receiver in Figure 8.2(a), there are several contributors to the noise at the input of the LNA. The main contributor is thermal noise coming from the antenna, which is known to be kTB, where k is the Boltzmann constant, T is the temperature in Kelvin, and B is the signal bandwidth. Accordingly, the thermal noise power in the signal band can be written as:

$$N_{thermal}[\text{dBm}] = 10\log(kTB) = 10log(kT)[\text{dBm/Hz}] + 10\log(B)[\text{dBHz}]. \quad (8.2)$$

At room temperature ($T = 300°\text{K}$), we can write:

$$N_{thermal}[\text{dBm}] = -174[\text{dBm/Hz}] + 10\log(B)[\text{dBHz}]. \quad (8.3)$$

The other contributor to the input noise is the LNA itself. As we mentioned earlier, the LNA is characterized by its NF, which quantifies the added noise in dB. As a result, the minimum equivalent noise at the input (input-referred noise) of the LNA can be written as:

$$\begin{aligned} N_{floor}[\text{dBm}] &= N_{thermal}[\text{dBm}] + NF[\text{dB}] \\ &= -174[\text{dBm/Hz}] + 10\log(B)[\text{dBHz}] + NF[\text{dB}]. \end{aligned} \quad (8.4)$$

If we require the minimum detectable signal to be an SNR_{min} above this noise level, then we can write an expression for the receiver sensitivity as follows:

$$RX_{sensitivity}[\text{dBm}] = N_{floor}[\text{dBm}] + SNR_{min}[\text{dB}]. \quad (8.5)$$

Figure 8.4 depicts the relationship between the thermal noise, the noise floor, and the RX sensitivity.

8.2 Examples of Phase Noise Requirements for the Transmitter VCO

In this section, we study two wireless standards, i.e., GSM and WCDMA, and determine the phase noise requirements of their transmitter VCOs. Figure 8.5 shows the frequency band associated with the P-GSM-900 standard [100]. In this standard, the TX band

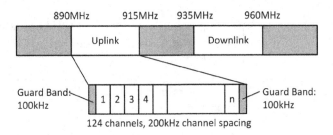

Figure 8.5 Primary GSM (P-GSM) standard spectrum.

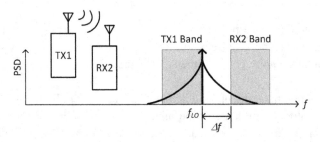

Figure 8.6 The local oscillator at the rightmost channel in the TX band leaks its phase noise to the leftmost channel of the RX band.

(mobile to base station), also known as uplink, is specified as $890 - 915\text{MHz}$, whereas the RX band (base station to mobile), also known as downlink, is specified as $935 - 960\text{MHz}$. With a channel spacing of 200kHz in each of the TX and RX bands, P-GSM-900 provides a total of 124 channels in each of its TX and RX bands with 100kHz of guard band on each side. A 20MHz gap between the TX and RX bands acts as a further guard band to ensure the phase noise of the TX signal from a handset does not spill too much into the RX band of an adjacent handset. We will determine the maximum phase noise at the TX that can be tolerated at the RX in this section.

Figure 8.6 shows the VCO phase noise associated with the rightmost channel in the TX band and how it spills into the leftmost channel in the RX band of an adjacent mobile set.

To determine the maximum phase noise of the TX that can be tolerated at the RX, and using the definition of phase noise in Equation 3.25 for a frequency offset of Δf, we can write:

$$\mathcal{L}_{GSM}(\Delta f) = \frac{S'(f_{LO} + \Delta f)}{P_{TX}}. \tag{8.6}$$

If the maximum noise power allowed in the receiver channel is specified as N_{RXmax}, we can write:

$$\mathcal{L}_{GSM}(\Delta f) \leq \frac{N_{RXmax}}{B \cdot P_{TX}}. \tag{8.7}$$

Expressing all quantities in dB, we can write:

$$\mathcal{L}_{GSM}(\Delta f)[\text{dBc/Hz}] \leq N_{RXmax}[\text{dBm}] - P_{TX}[\text{dBm}] - 10\log(B)[\text{dBHz}]. \tag{8.8}$$

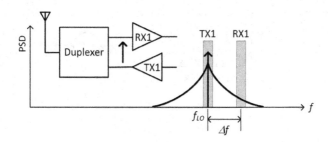

Figure 8.7 The phase noise of the TX channel of a WCDMA device leaks into the RX channel of the same device.

If we evaluate this equation with $\Delta f = 20$MHz, $N_{RXmax} = -79$dBm, $P_{TX} = 33$dBm, and $B = 100$kHz, we will arrive at the following requirement:

$$\mathcal{L}_{GSM}(20\text{MHz}) \leq -162\text{dBc/Hz}. \tag{8.9}$$

As a second example, let us find the TX phase noise requirement in a WCDMA standard [100]. As shown in Figure 8.7, in this standard, the phase noise of the TX of one handset may actually interfere with the RX part of the same handset. Although the duplexer is expected to block any of the handset's TX power from leaking into the handset's RX, in reality, the duplexer can only attenuate the TX power by about 45dB. If we consider a maximum TX power of 26dBm, the duplexer is able to bring this power down to $26\text{dBm} - 45\text{dB} = -19$dBm. Δf can be any value between 30MHz and 400MHz depending on the specific band. In this case, Equation 8.8 will result in $\mathcal{L}_{WCDMA}(\Delta f) \leq -161$dBc/Hz [101].

8.3 Reciprocal Mixing at the Receiver

An ideal RX mixer, as shown in Figure 8.8, multiplies a narrowband RF signal around f_{RF} by a tone at f_{LO} to reproduce the narrowband signal around an IF frequency where ($f_{IF} \ll f_{RF}$). The basic idea is that the content centered around f_{IF} is much easier to process than the original content centered around f_{RF}.

The tone at f_{LO}, which is generated using a local oscillator (LO), inevitably contains phase noise, as discussed in detail in Chapter 4. Figure 8.9 shows a tone at f_{LO} and its accompanying phase noise. Consider now a blocker signal with sufficient strength at f_B. The blocker signal, when mixed with the phase noise component at frequency $f_B + f_{IF}$, produces content around f_{IF}, in a similar way to how the signal around f_{RF} is translated by the ideal tone at f_{LO} to around f_{IF}. The blocker signal, as discussed earlier, may belong to either the same band as that of the desired signal or to an adjacent band. In most narrowband RF receivers, an off-chip filter (SAW filter in Figure 8.2) is employed to attenuate the out-of-band blocker signal. In wideband RF receivers without a SAW filter, the blocker signal may accompany the desired signal to the mixer input and impose more constraints on LNA linearity and on the phase noise of the local VCO.

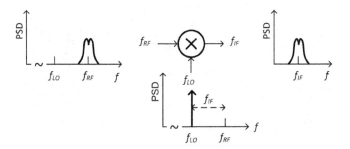

Figure 8.8 An ideal mixer shifts the content from f_{RF} to f_{IF}.

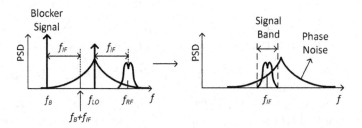

Figure 8.9 The LO phase noise is mixed with the blocker signal and produces in-band noise at f_{IF}.

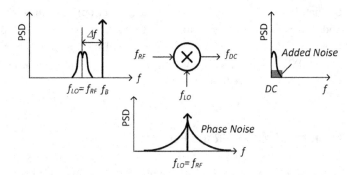

Figure 8.10 In a direct-conversion receiver, the LO phase noise is mixed with the blocker signal and produces in-band noise at DC.

The process in which the blocker creates more in-band noise through mixing with the phase noise of the VCO is called "reciprocal" mixing because the blocker signal at f_B plays the role of a local oscillator and the phase noise of the local oscillator plays the role of the RF signal. Let us now quantify the effect of this reciprocal mixing on the SNR at the mixer output.

Without loss of generality, we assume $f_{IF} = 0$ in this analysis. This corresponds to what is known as direct-conversion receiver. Figure 8.10 shows the desired signal and a blocker at Δf away from f_{LO} as input to the mixer. If we assume the local oscillator has a phase noise characterized by $\mathcal{L}_{LO}(f)$ and the blocker has a total power of P_B[dBm], then the process of reciprocal mixing will add the following noise power to the signal band:

$$N_{RM} = P_B \cdot \mathcal{L}_{LO}(\Delta f) \cdot B \tag{8.10}$$

where N_{RM} denotes the in-band noise power due to reciprocal mixing and B is the signal bandwidth. Expressing this noise power in dBm, we can write:

$$N_{RM}[\text{dBm}] = P_B[\text{dBm}] + \mathcal{L}_{LO}(\Delta f)[\text{dBc/Hz}] + 10\log(B)[\text{dBHz}]. \tag{8.11}$$

If we assume this noise is much larger than the thermal noise, which is usually the case given large blocker power, we can write an expression for the SNR:

$$SNR[\text{dB}] = P_C[\text{dBm}] - P_B[\text{dBm}] - \mathcal{L}_{LO}(\Delta f)[\text{dBc/Hz}] - 10\log(B)[\text{dBHz}] \tag{8.12}$$

where P_C is the carrier power associated with the desired signal. If we denote the minimum required SNR by the standard with SNR_{min}, then we have the following requirement for the phase noise:

$$\mathcal{L}_{LO}(\Delta f)[\text{dBc/Hz}] \leq P_C[\text{dBm}] - P_B[\text{dBm}] - 10\log(B)[\text{dBHz}] - SNR_{min}[\text{dB}]. \tag{8.13}$$

Example: Figure 8.11 shows the GSM receiver requirement for the blocker power level. Given the GSM standard requires a minimum SNR of 9dB and a sensitivity of −99dBm, i.e., a carrier level of −99dBm, calculate the LO phase noise at 600kHz and 3MHz offset from the carrier. Assume the blocker signals have the maximum allowable power levels as shown in the figure. Assume a channel bandwidth of 200kHz.

Solution: The blocker power can be as much as −43dBm at 600kHz away from the carrier. Therefore, using Equation 8.13, we can write:

$$\mathcal{L}_{LO}(\Delta f = 600kHz) \leq -99 - (-43) - 10\log(200k) - 9 = -118[\text{dBc/Hz}]. \tag{8.14}$$

Similary, the blocker power can be as much as −23dBm at 3MHz away from the carrier. Therefore, we can calculate:

$$\mathcal{L}_{LO}(\Delta f = 3MHz) \leq -99 - (-23) - 10\log(200k) - 9 = -138[\text{dBc/Hz}]. \tag{8.15}$$

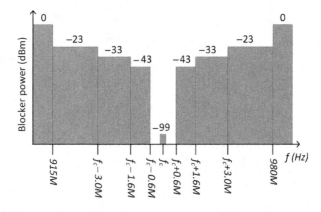

Figure 8.11 Blocker mask for GSM: the receiver is required to tolerate an increasing level of blocker power as the blocker moves farther away from the desired signal frequency.

Finally, to find out the increase in noise power due to reciprocal mixing, we need to compare this noise power with the thermal noise power, which is $-174[\text{dBm/Hz}] + 10\log(B)[\text{dBHz}]$. Subtracting the two in dB will provide us with the blocker NF:

$$BlockerNF[\text{dB}] = P_B[\text{dBm}] + 174[\text{dBm}] + \mathcal{L}_{LO}(\Delta f)[\text{dBc/Hz}]. \qquad (8.16)$$

The blocker NF can be reduced if we can reduce the blocker power (P_B) or the VCO's phase noise. As mentioned earlier, one way to reduce the blocker power is to simply try to filter it using a SAW filter. However, in the absence of a SAW filter, we need to design the local oscillator with reduced phase noise. Equation 8.16 states that every dB increase in blocker power must be compensated by a dB decrease in the phase noise, or else we will be compromising the SNR of the design.

Since the phase noise spectrum is usually symmetric around the carrier frequency, as depicted in Figure 8.10, it is possible to take advantage of this symmetry and cancel the reciprocal mixing caused by the phase noise. Details of this technique are described in [102].

9   Advanced Concepts on Jitter and Phase Noise

This chapter deals with some advanced concepts connected to jitter and phase noise which are not strictly necessary to understand the rest of the book. The handling of these concepts requires, in some cases, a more complex mathematical approach, but in our view is important for those readers who want to gain a deeper insight.

9.1   A General Method to Convert Phase Noise to Jitter

In this section we will illustrate a general method that can be employed in many cases to calculate jitter, or other timing parameters, from the phase noise. This method borrows some basic concepts from discrete time signal processing, which can be found in, e.g., [72]. Assume that a given timing parameter $\mathbf{j}$ is such that a linear combination of its values over the clock periods $k, \ldots, k - n$ can be expressed as a linear combination of the absolute jitter $\mathbf{a}$ over the clock periods $k, \ldots, k - m$:

$$\alpha_0 \mathbf{j}_k + \alpha_1 \mathbf{j}_{k-1} + \cdots + \alpha_n \mathbf{j}_{k-n} = \beta_0 \mathbf{a}_k + \beta_1 \mathbf{a}_{k-1} + \cdots + \beta_m \mathbf{a}_{k-m}. \tag{9.1}$$

As an example, the period jitter $\mathbf{p}_{k-1}$ can be expressed as $\mathbf{a}_k - \mathbf{a}_{k-1}$ (see Equation 2.9). By taking the z-transform of both members of the previous equation, and indicating with $\mathbf{J}(z)$ and $\mathbf{A}(z)$ the z-transform of $\mathbf{j}_k$ and $\mathbf{a}_k$ respectively:

$$(\alpha_0 + \alpha_1 z^{-1} + \cdots + \alpha_n z^{-n})\mathbf{J}(z) = (\beta_0 + \beta_1 z^{-1} + \cdots + \beta_m z^{-m})\mathbf{A}(z). \tag{9.2}$$

Therefore, in general the discrete time signal $\mathbf{j}_k$ can be obtained from $\mathbf{a}_k$ via a discrete time filter with frequency response $H(z)$:

$$\mathbf{J}(z) = H(z)\mathbf{A}(z) \tag{9.3}$$

where, in the specific case above, $H(z)$ is defined as:

$$H(z) := \frac{\sum_{i=0}^{m} \beta_i z^{-i}}{\sum_{i=0}^{n} \alpha_i z^{-i}}. \tag{9.4}$$

Knowing the general relation between absolute jitter and excess phase $\mathbf{a}_k = -\varphi_k/\omega_0$ (see Equation 3.5), Equation 9.3 can be expressed as:

$$\mathbf{J}(z) = \frac{-H(z)}{\omega_0}\Phi(z) \tag{9.5}$$

where $\Phi(z)$ is the z-transform of the sequence φ_k. This relation tells us that $\mathbf{j}$ is obtained from φ via a filter with frequency response $-H(f)/\omega_0$, where the translation from z-domain to frequency domain is obtained by replacing z^{-1} with $e^{-j2\pi fT}$, with T the nominal clock period. Using basic linear system theory (see Appendix A.2.10), the relation between the PSD of $\mathbf{j}$ and the PSD of φ, which is also the phase noise $\mathcal{L}(f)$, can then be found:

$$S_{\mathbf{j}}(f) = \left| \frac{H(f)}{\omega_0} \right|^2 S_\varphi(f) = \left| \frac{H(f)}{\omega_0} \right|^2 \mathcal{L}(f). \tag{9.6}$$

From here it is easy to obtain the variance of the timing parameter $\mathbf{j}$ as:

$$\sigma_{\mathbf{j}}^2 = 2 \int_0^{+\infty} \left| \frac{H(f)}{\omega_0} \right|^2 \mathcal{L}(f) \, df. \tag{9.7}$$

EXAMPLE 15   In this example we will apply the method above to the computation the period jitter from the phase noise. Starting from the definition of period jitter (see Equation 2.9) here reported:

$$\mathbf{p}_{k-1} = \mathbf{a}_k - \mathbf{a}_{k-1} \tag{9.8}$$

and taking the z-transform of both members we find:

$$z^{-1}\mathbf{P}(z) = (1 - z^{-1})\mathbf{A}(z) \tag{9.9}$$

so that for this case:

$$H(z) = \frac{1 - z^{-1}}{z^{-1}}. \tag{9.10}$$

Using Equation 9.6, we obtain:

$$S_{\mathbf{p}}(f) = \left| \frac{1 - e^{-j2\pi fT}}{\omega_0 e^{-j2\pi fT}} \right|^2 \mathcal{L}(f) = \frac{4 \sin^2(\pi fT)}{\omega_0^2} \mathcal{L}(f) \tag{9.11}$$

and finally, using Equation 9.7, the variance of the period jitter as function of the phase noise:

$$\sigma_{\mathbf{p}}^2 = \frac{8}{\omega_0^2} \int_0^{+\infty} \sin^2(\pi fT)\mathcal{L}(f) \, df \tag{9.12}$$

which is the same as 3.49 for $N = 1$, obtained in a different way.

The reader is invited to extend the example above to derive Equation 3.49, which converts phase noise into N-period jitter. In Section 9.3.4 this method will be applied to derive the link between Allan Deviation and phase noise.

9.2   Confidence Intervals of Statistical Parameters

In Section 2.2, the statistical parameters of jitter were presented and analyzed, but no mention was made about their accuracy. This important aspect is the topic of this section.

First of all, it is important to understand that the quantities $\hat{\mu}_{\mathbf{j}}$ and $\hat{\sigma}_{\mathbf{j}}^2$ as calculated in Equations 2.30 and 2.32 only deliver an *estimate* of mean and variance. Due to the stochastic nature of the data, the same computation applied to data coming from the same system but measured at different times will give different results, even if the system is time-invariant. In other words, the quantities $\hat{\mu}_j$ and $\hat{\sigma}_j^2$ should be treated as random variables on their own.

Additionally, the accuracy of the estimation depends on the sample size n. It is natural to expect an estimate of the variance based on a thousand samples to be much closer to the real variance than an estimate based on five samples. Figure 9.1 shows the results of the repeated computation of the sample variance using Equation 2.32 on data coming from a Gaussian distribution with a variance equal to 1 for several sample sizes n. As can be seen, the larger the n, the smaller the spread of the sample variance around the real value.

The central questions that must be answered when estimating mean and variance of jitter are, therefore: if the number of samples n is given and cannot be changed, how close is the result of the estimation to the real value? Or, if the sample size n can be chosen, how large should it be so that the error in the estimation is lower than a given limit?

These questions constitute the core of so-called *parameter estimation theory* in statistics. In this section, we will use the concepts from this branch of statistics without going into the details of the theory. The interested reader can refer to, e.g., [103], [104], [105], or [13] for a deeper analysis. The content of this section is not a prerequisite for understanding the remainder of the book.

The questions above can be formalized by introducing the concept of *confidence interval*. If θ is the parameter to be estimated, in our case, either the mean or the variance, the goal is to find two values θ_L and θ_U such that the probability that the real value θ falls within the interval $[\theta_L, \theta_U]$ is equal to a given number $1 - \alpha$:

$$P[\theta_L \leq \theta \leq \theta_U] = 1 - \alpha. \tag{9.13}$$

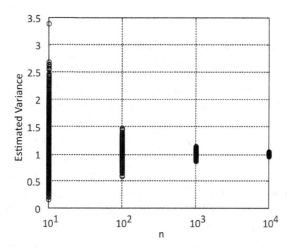

Figure 9.1 Estimated (sample) variance from a Gaussian population with zero mean and variance equal to 1, versus sample size n. For each sample size n, the results over 1000 trials are displayed.

The interval $[\theta_L, \theta_U]$ is called the $100(1 - \alpha)\%$ confidence interval, and it has the following meaning. If the computation of the interval $[\theta_L, \theta_U]$ is repeated many times on different data sets, in $100(1 - \alpha)\%$ of the cases the real value of the parameter will be contained within the computed confidence interval. For instance, if $\alpha = 0.01$, we have a 99% confidence interval; that is, in 99% of the cases the real value of θ will be within the computed limits. For some of us, it could be particularly disillusioning to realize that we cannot know the value of the parameter with absolute certainty, but fighting against this statistical truth will probably not lead us anywhere. So we'd better accept it, for now.

Since the computation of the confidence intervals is heavily based on the Gaussian distribution, it is useful to review some facts regarding the probability of a Gaussian random variable before proceeding. Given a random variable X having a Gaussian distribution with mean μ and variance σ^2, the probability $P(Q)$ that X assumes values larger than $\mu + Q\sigma$, with Q a real number, is given by:

$$P(Q) := P[X \geq \mu + Q\sigma] = \int_{\mu+Q\sigma}^{+\infty} \frac{1}{\sigma\sqrt{2\pi}} \exp\left(\frac{-(t - \mu)^2}{2\sigma^2}\right) dt. \qquad (9.14)$$

With the change of variable $v = (t - \mu)/(\sigma\sqrt{2})$ this expression can be rewritten as:

$$P(Q) = \frac{1}{\sqrt{\pi}} \int_{Q/\sqrt{2}}^{+\infty} e^{-v^2} dv = \frac{1}{2}\mathrm{erfc}\left(\frac{Q}{\sqrt{2}}\right) \qquad (9.15)$$

where $\mathrm{erfc}(x)$ is the complementary error function defined as:

$$\mathrm{erfc}(x) := \frac{2}{\sqrt{\pi}} \int_x^{+\infty} e^{-v^2} dv \qquad (9.16)$$

which can be computed with the use of the most common mathematical programs or spreadsheets. Table 9.1 shows the values of $P(Q)$ for several values of Q. In the following a Gaussian random variable X with mean μ and variance σ^2 will be indicated by $X \in \mathcal{N}(\mu, \sigma^2)$.

9.2.1 Confidence Interval on the Jitter Mean

In this section we will derive an expression for the confidence interval of the estimation of mean based on the sample mean (Equation 2.30). From Equation 2.30, if the jitter samples j_k have a Gaussian distribution $\mathcal{N}(\mu, \sigma^2)$, and are independent from each other, it is well known that $\hat{\mu}_j \in \mathcal{N}(\mu, \sigma^2/n)$, since the sum of independent Gaussian random variables is still a Gaussian random variable. If the jitter samples are non-Gaussian, the central limit theorem states that, under certain conditions, the distribution of $\hat{\mu}_j$ converges to a normal distribution $\mathcal{N}(\mu, \sigma^2/n)$ for increasing n.

In practical situation, if $n \geq 30$ the distribution approximates a real Gaussian distribution to a very high degree (see Figure 9.2). Luckily, in the context of jitter estimation, the number of samples available in practice is far larger than 30, and it can be assumed that:

$$\hat{\mu}_j \in \mathcal{N}\left(\mu, \frac{\sigma^2}{n}\right). \qquad (9.17)$$

Table 9.1 Probabilities that $X \geq \mu + Q\sigma$ for $X \in \mathcal{N}(\mu, \sigma^2)$.

Q	$P(Q)$
1	$1.58 \cdot 10^{-1}$
2	$2.27 \cdot 10^{-2}$
2.33	$1.00 \cdot 10^{-2}$
3	$1.35 \cdot 10^{-3}$
3.09	$1.00 \cdot 10^{-3}$
4	$3.17 \cdot 10^{-5}$
4.75	$1.00 \cdot 10^{-6}$
5	$2.87 \cdot 10^{-7}$
6	$1.00 \cdot 10^{-9}$
7	$1.28 \cdot 10^{-12}$
7.03	$1.00 \cdot 10^{-12}$
7.94	$1.00 \cdot 10^{-15}$
8	$6.22 \cdot 10^{-16}$
9	$1.13 \cdot 10^{-19}$
10	$7.62 \cdot 10^{-24}$

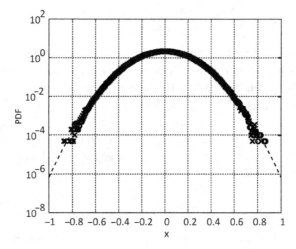

Figure 9.2 PDF of the sample mean with $n = 30$ from non-Gaussian populations (circles = uniform, crosses = sinusoidal) having variance 1 compared to the Gaussian PDF with variance $1/\sqrt{30} = 0.18$ (dashed line).

From this result it is easy to find the confidence interval for the sample mean. Indeed, from Equation 9.14

$$P\left[\hat{\mu}_j \geq \mu + Q\frac{\sigma}{\sqrt{n}}\right] = P(Q) \tag{9.18}$$

so that:

$$P\left[\mu + Q\,\frac{\sigma}{\sqrt{n}} \le \hat{\mu}_j \le \mu + Q\,\frac{\sigma}{\sqrt{n}}\right] = 1 - 2P(Q) \tag{9.19}$$

or, in term of the absolute estimation error $|\hat{\mu} - \mu|$:

$$P\left[|\hat{\mu}_j - \mu| \le Q\,\frac{\sigma}{\sqrt{n}}\right] = 1 - 2P(Q). \tag{9.20}$$

It can be seen that the confidence interval depends on the sample size n. The larger the n, the smaller the confidence interval for a given confidence level $1 - 2P(Q)$. This expression allows for the calculation of the minimum number of samples n needed to achieve the $1 - 2P(Q)$ confidence level that the absolute error on the estimated mean ($|\hat{\mu}_j - \mu|$) is lower than a given value ϵ:

$$n \ge \left(\frac{Q\sigma}{\epsilon}\right)^2. \tag{9.21}$$

EXAMPLE 16   Assume it is known that the clock under investigation has a jitter RMS value of 10ps, and we want to estimate the jitter mean with an absolute error lower than 1ps and a confidence level of 99.7%. How many jitter samples are needed? Based on Table 9.1, for this confidence level $Q = 3$, so that $n \ge (3 \cdot 10/1)^2 = 900$. It can be seen that, for the estimation of the mean, a number of samples as low as 1000 already yields a very good confidence estimation. Note that, for a confidence level of 95%, $Q = 2$, so that only 400 samples would suffice.

In the previous discussion, it has been tacitly assumed that the RMS value σ of the jitter was known. In a realistic situation, though, if the mean is unknown, it not very probable that the RMS will be known. The most natural choice in this case is to use the estimated RMS value $\hat{\sigma}_j$ from Equation 2.32 instead of σ. At this point, statistical theory pulls in the concept of the Student's T-distribution and things get slightly more complicated. However, again luckily, if the number of samples n is greater than about 30, the T-distribution can be very well approximated with a Gaussian one and the result derived above holds its validity when σ is replaced by $\hat{\sigma}_j$:

$$P\left[|\hat{\mu}_j - \mu| \le Q\,\frac{\hat{\sigma}_j}{\sqrt{n}}\right] = 1 - 2P(Q). \tag{9.22}$$

However, the determination of the minimum sample size n is now not as straightforward as in the case of a known σ, since $\hat{\sigma}_j$ depends also on n. Therefore, a trial and error procedure is normally needed.

9.2.2   Confidence Interval on the Jitter Variance and Standard Deviation

In most cases, the parameter of interest is the RMS value or the variance of the jitter and not its mean value. It is thus important to know the confidence interval on the estimation of these quantities when using the sample variance estimator $\hat{\sigma}_j^2$ reported in Expression 2.32.

From mathematical statistics (see, e.g., [104]) it is known that, if the jitter samples $\mathbf{j}_k \in \mathcal{N}(\mu, \sigma^2)$, then the random variable $(n-1)\hat{\sigma}_j^2/\sigma^2$ has a χ^2 distribution with $n - 1$

degrees of freedom. The χ^2 is a special distribution used extensively in statistical inference and can be defined with the help of the Gamma function. However, and luckily enough, it has been shown that, for a large number of degrees of freedom n, the χ^2 distribution converges to $\mathcal{N}(n, 2n)$. The approximation is already very good for $n = 30$. Using this fact, and approximating $n - 1$ with n, it follows that:

$$\hat{\sigma}_j^2 \in \mathcal{N}(\sigma^2, 2\sigma^4/n). \tag{9.23}$$

This expression means that, when estimating the variance of jitter with the sample variance, the result will be, on average, the real variance σ^2, but it will be spread around this average value with the shape of a Gaussian distribution with an RMS value $\sigma^2\sqrt{(2/n)}$. Note that the *RMS* value of the sample variance is proportional to the *variance* of the original random variable.

Now it is quite straightforward to calculate the confidence interval. From Equation 9.23:

$$P\left[\sigma^2 - Q\sigma^2\sqrt{\frac{2}{n}} \le \hat{\sigma}_j^2 \le \sigma^2 + Q\sigma^2\sqrt{\frac{2}{n}}\right] = 1 - 2P(Q). \tag{9.24}$$

After some easy algebra, the previous expression can be reformulated as:

$$P\left[\left|\frac{\hat{\sigma}_j^2}{\sigma^2} - 1\right| \le Q\sqrt{\frac{2}{n}}\right] = 1 - 2P(Q). \tag{9.25}$$

Note that the expression $\hat{\sigma}_j^2/\sigma^2 - 1$ is the relative error of the estimation of the variance. The confidence interval for the estimation of the RMS value can be derived from Equation 9.25, observing that, if $\hat{\sigma}_j$ is close to σ, then $\hat{\sigma}_j^2/\sigma^2 - 1 \approx 2(\hat{\sigma}_j/\sigma - 1)$, and $\hat{\sigma}_j/\sigma - 1$ is the relative error of the RMS value. Finally, the confidence interval of the relative error of the estimation of the RMS value is:

$$P\left[\left|\frac{\hat{\sigma}_j}{\sigma} - 1\right| \le Q\sqrt{\frac{1}{2n}}\right] = 1 - 2P(Q). \tag{9.26}$$

From this formula, the minimum number of samples n needed to estimate the RMS with a given confidence level can be derived immediately. Assuming that the allowed relative error of the estimation is ϵ_r, the minimum number of samples n corresponding to a confidence level of $1 - 2P(Q)$ is given by:

$$n \ge \frac{1}{2}\left(\frac{Q}{\epsilon_r}\right)^2. \tag{9.27}$$

Figure 9.3 graphically illustrates Equation 9.27 for three levels of confidence: 68% ($Q = 1$), 95% ($Q = 2$) and 99.7% ($Q = 3$).

EXAMPLE 17 How many samples are needed to estimate the RMS value of a Gaussian jitter with a relative error of 1% and a 99.7% confidence level? In this case $\epsilon_r = 0.01$ and $Q = 3$, so that, from Equation 9.27, n should be larger than 45,000. Note that for a confidence level of 95% and the same relative error, the sample size reduces to 20,000.

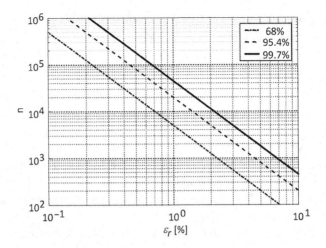

Figure 9.3 Sample size n versus relative error ϵ_r for the estimation of the RMS value of a Gaussian random variable for three levels of confidence.

The expressions above are valid if the jitter samples $\mathbf{j}_k$ have a Gaussian distribution. What if this condition is not satisfied? In such cases, things are not as straightforward as before, but we can contain the increase of complexity by introducing just one additional coefficient. By virtue of the central limit theorem for large n, the distribution of the sample variance tends again to a Gaussian distribution, with the only difference being that in the non-Gaussian case, the variance is not $2\sigma^4/n$, but rather $\sigma^4(2 + \gamma_2)/n$:

$$\hat{\sigma}_{\mathbf{j}}^2 \in \mathcal{N}\left(\sigma^2, \frac{\sigma^4}{n}(2 + \gamma_2)\right). \tag{9.28}$$

The coefficient γ_2 is called the *excess kurtosis* and, for a random variable X with mean μ and variance σ^2, is defined as:

$$\gamma_2 := \frac{\mathrm{E}\left[(X - \mu)^4\right]}{\sigma^4} - 3. \tag{9.29}$$

It is basically a measure of how close to a Gaussian the distribution of X is. Indeed if X is Gaussian then $\gamma_2 = 0$ and Equation 9.28 reduces to Equation 9.23. It must be noted that also in the case of non Gaussian distributions, the sample variance converges to the real variance, and the excess kurtosis influences only the confidence interval of the estimation. Repeating the steps followed above for the Gaussian case, the confidence interval for the estimated RMS value for non-Gaussian jitter distributions can be written as:

$$P\left[\left|\frac{\hat{\sigma}_{\mathbf{j}}}{\sigma} - 1\right| \leq \frac{Q}{2}\sqrt{\frac{2 + \gamma_2}{n}}\right] = 1 - 2P(Q) \tag{9.30}$$

and the minimum sample size n for a given confidence level $1 - 2P(Q)$ as:

$$n \geq \left(\frac{1}{2} + \frac{\gamma_2}{4}\right)\left(\frac{Q}{\epsilon_r}\right)^2. \tag{9.31}$$

Unfortunately, this approach relies on the fact that we know the value of γ_2; that is, we know the shape of the underlying jitter distribution. In most cases, though, this is not true, and the excess kurtosis becomes another coefficient that we have to estimate using Equation 9.29, with its own confidence interval (the order-zero estimation can be obtained by replacing the expectation operator with the average over the samples). As can be understood, an exact solution to this problem is not elementary and goes far beyond the scope of this book.

9.2.3 Confidence Interval on the Jitter Peak and Peak-Peak Values

Calculating peak and peak-peak values, as shown in Equations 2.33, 2.34, and 2.35, is very straightforward, but what confidence level can be obtained with this procedure? In other words, once ρ^-, ρ^+, and $\rho\rho$ are found, what is the probability that a jitter sample will be larger that ρ^+, or smaller than ρ^-, or outside the range $\rho\rho$?

This question is part of a branch of mathematics called *order statistics* developed in the 1940s and applied to quality control in industrial processes. In particular, a quite remarkable result states that the probability of finding a jitter sample outside the limits calculated as before depends only on the sample size n and not on the shape of the distribution, as long as it can be assumed to be continuous. The proof of this apparently paradoxical fact is beyond the scope of this book, and the interested reader is referred to [106] or [107].

The question now becomes slightly different. Given that these probabilities depend only on n, what is the minimum n such that there is a high confidence level, e.g., $1-P(Q)$, with $P(Q)$ as reported in Table 9.1, that the probability of finding a jitter sample beyond the peak limits is lower than a given bound β? Expressed in mathematical terms, calling u and v the probabilities of finding a jitter sample larger than ρ^+ or lower than ρ^-, respectively, we want to determine n such that:

$$P[u \leq \beta] \geq 1 - P(Q) \tag{9.32}$$

for the peak positive,

$$P[v \leq \beta] \geq 1 - P(Q) \tag{9.33}$$

for the peak negative, and

$$P[u + v \leq \beta] \geq 1 - P(Q) \tag{9.34}$$

for the peak-peak value. Omitting the mathematics (see [107]), the result is:

$$n \geq \frac{\ln(P(Q))}{\ln(1 - \beta)} \tag{9.35}$$

for the peak positive and the peak negative cases. The result is plotted in Figure 9.4 for three levels of confidence (68%, 95.4%, and 99.7%). For the peak-peak value, n is the result of the following transcendental equation:

$$n(1 - \beta)^{n-1} - (n - 1)(1 - \beta)^n = P(Q) \tag{9.36}$$

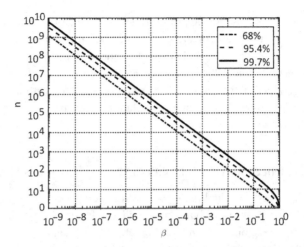

Figure 9.4 Estimation of peak value: sample size n versus probability bound β for three confidence levels.

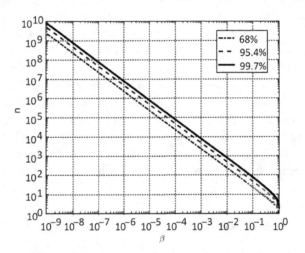

Figure 9.5 Estimation of peak-peak value: sample size n versus probability bound β for three confidence levels.

which cannot be solved analytically, but is plotted in Figure 9.5 for the same three levels of confidence.

It is instructive to find an approximate expression for Equation 9.36 in the case of a 99.7% confidence interval ($P(Q) = 0.003$). Considering the Taylor expansion of the logarithm for small values of β, the result is:

$$n \geq \frac{5.81}{\beta}. \tag{9.37}$$

For instance, if one wants to be 99.7% sure that at most 1/1000 of the jitter samples are larger than the peak value, n should be larger than 5810. Comparing Figures 9.4 and 9.5, it can be seen that the same level of confidence for the same value of β requires roughly double the sample size in the peak-peak case than in the case of peak value. This also makes intuitive sense: having a proportion β outside of the $\rho\rho$ range can be equivalent to having $\beta/2$ larger than ρ^+ and $\beta/2$ smaller than ρ^-, leading to the doubling of the number of points, as shown by Equation 9.37.

9.3   Estimators for Frequency Stability

Many modern applications rely on the availability of timing sources where frequency is accurate, i.e., its average value is equal to the nominal one, and very stable, i.e., not changing too much over time. Notable examples are the GPS system and digital communication networks extending across very large geographic distances. In the case of GPS, the phase differences between clock generators in different satellites and the receiver are used to determine distance, and thus the position of the receiver itself. A key point for the system to work is that the clocks should be very stable and accurate, otherwise the information on distance extracted by comparing phases will be corrupted. In digital communication networks, frequency stability over the long term at different nodes geographically far apart is a key element in guaranteeing that data are not lost due to buffer overflow.

In the 1950s and 1960s a lot of effort was put into devising clock sources offering the kind of stability needed by such applications, resulting in the development of timing sources which exploit basic properties of matter to generate periodic signals with controlled frequencies. Atomic clocks based on the energy transitions of cesium or rubidium, and the hydrogen maser are some of these.

In parallel with the development of precise timing sources, there was a clear necessity of establishing criteria to characterize the frequency stability over time of those different sources. Two factors prompted this development: the first was the need to compare the performances of clock sources, and the second was the need for a system to classify the degrees of frequency accuracy required by different applications.

In the late 1960s the metrology community converged on a few well-defined quantities, which proved to be quite useful and are still in use today. The Allan Deviation is probably the most well known.

Even though these quantities are not directly connected to the design and analysis of integrated systems, they are quite relevant for many other applications in the field of electronic and information engineering. This section aims to give an overview of the subject; a thorough handling would take a book by itself. The interested reader is referred to [108], [109], and [110] for an in-depth examination of these issues.

This section will start with some basic concepts dealing with the characterization of frequency stability and then focus on the Allan Deviation and on its relation to jitter. At the end, the concepts of Modified Allan Deviation and Time Deviation will be briefly explained, but not discussed in detail.

9.3.1   Basic Concepts

The main question to address is how to determine the frequency stability of a given clock source over a given observation period τ. The starting point for an analysis in the time domain is the simple clock model described by Equation 2.1. The quantity to characterize is the deviation over time of the instantaneous frequency of the clock signal $f(t) = f_0 + \dot{\varphi}(t)/(2\pi)$ from the average nominal frequency f_0, $\Delta f(t) = f(t) - f_0$. It makes sense to consider the frequency deviation normalized to the nominal frequency, so that historically a new quantity $y(t)$ was defined as:

$$y(t) := \frac{\Delta f(t)}{f_0} = \frac{f(t) - f_0}{f_0}. \tag{9.38}$$

This is essentially the frequency error over time normalized to the nominal frequency. Note that, for an ideal clock, $y(t)$ is identical to zero.

In order to evaluate the stability over an observation period τ, the approach used by the standardization bodies considered the average values of $y(t)$ over periods of time spanning τ seconds. It must be noted that, depending on the application, the observation period τ can range from milliseconds to days or even years. Figure 9.6 shows an example of $y(t)$ together with its average values (indicated as $\bar{y}_i$) calculated over consecutive periods τ.

It would seem sensible to characterize the stability of the clock source by simply taking the variance of the $\bar{y}_i$. This is indeed one way of doing it, leading to a quantity which has been called *true variance* and indicated by $I^2(\tau) = \mathrm{E}\left[\bar{y}_i^2\right]$. Unfortunately, this quantity turns out to be of no practical use, since, for all real clock sources, it approaches infinity for increasing values of τ, due to the intrinsic random walk nature of the frequency error for all real oscillators, so that its variance diverges with time.

For this reason, the Allan Deviation (named after D. W. Allan) was taken as the basic metrological quantity for frequency stability.

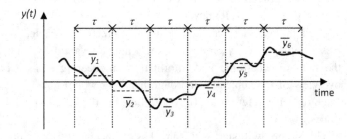

Figure 9.6 Normalized frequency deviation $y(t)$ versus time. The $\bar{y}_i$ indicate the average values of $y(t)$ across a time span τ.

9.3.2   Allan Deviation (ADEV)

To get rid of the numerical difficulties connected with the true variance, other approaches were considered, based on the relative comparison of the $\bar{y}_i$, rather than on their absolute values. One of them is built on the difference between two consecutive values of $\bar{y}_i$, by defining the quantity Δy_i as:

$$\Delta y_i := \frac{\bar{y}_{i+1} - \bar{y}_i}{\sqrt{2}}. \tag{9.39}$$

The variance of Δy is called the Allan Variance (AVAR) $\sigma^2_{\Delta y}(\tau)$:

$$\sigma^2_{\Delta y}(\tau) := \mathrm{E}\left[(\Delta y_i)^2\right] = \frac{1}{2}\mathrm{E}\left[(\bar{y}_{i+1} - \bar{y}_i)^2\right] \tag{9.40}$$

and its square root $\sigma_{\Delta y}(\tau)$ is called the Allan Deviation (ADEV). Since the values of $\bar{y}_i$ depend on the observation interval τ, the ADEV is also a function of τ. Note that, in the literature, the ADEV is normally indicated by $\sigma_y(\tau)$. Nevertheless, since it is based on the difference of consecutive $\bar{y}_i$, in this book we prefer to use the notation $\sigma_{\Delta y}(\tau)$, to avoid possible confusion with the standard deviation of $\bar{y}_i$ itself. The factor $\sqrt{2}$ in the denominator of Equation 9.39 is chosen so that the value of ADEV is equal to the standard deviation of $\bar{y}_i$ for a noise process in which consecutive values of $\bar{y}_i$ are uncorrelated. Due to the difference operation in Equation 9.39, the ADEV turns out to be well defined and not diverging for any kind of noise produced by real oscillators. Given a series of m average frequency measurements, a widely used estimator for the AVAR is given by:

$$\sigma^2_{\Delta y}(\tau) = \frac{1}{2(m-1)}\sum_{i=1}^{m-1}(\bar{y}_{i+1} - \bar{y}_i)^2. \tag{9.41}$$

Figure 9.7 [47] shows the ADEV for some typical high-precision clock sources.

9.3.3   Relation Between Allan Deviation and Jitter

Since the ADEV is a characterization of frequency instability, and frequency instability manifests itself as jitter on the clock edges, it is natural to try to understand if there is a simple relation between ADEV and the jitter definitions given in this chapter. It turns out that there is, and it is simple indeed.

To derive it, let's first consider the very definition of $y(t)$ as given in Equation 9.38. If the frequency deviation Δf is small compared to f_0, the ratio of the two is equal to the clock period deviation, ΔT, divided by the nominal period, $T = 1/f_0$.[1] Under this hypothesis, which is very well satisfied for the case of high-accuracy clocks where the concept of ADEV is used, the $\bar{y}_i$ can be expressed as:

$$\bar{y}_i = -\frac{\overline{\Delta T}_i}{T} = -\frac{\overline{T}_i - T}{T} \tag{9.42}$$

---

[1] This follows from the fact that period and frequency are reciprocals of each other, $f = 1/T$. Differentiating f, we find $df = -dT/T^2$, so that $df/f = -dT/T$.

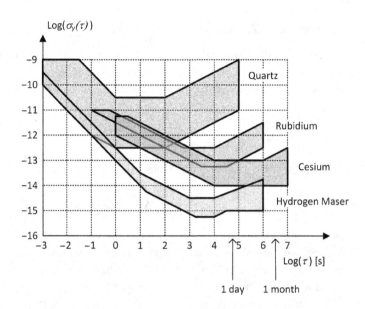

Figure 9.7 Allan deviation (ADEV) for some typical high-precision clock sources (reproduced with permission from Dr. John Vig[47]).

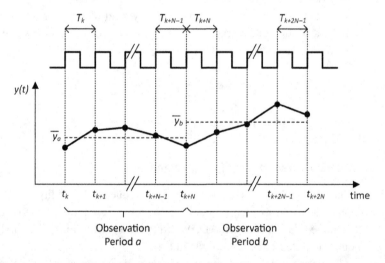

Figure 9.8 Computation of ADEV from jitter.

where T is the nominal period, $\overline{\Delta T_i}$ and $\overline{T_i}$ are the average period deviation and the average period respectively, across the i-th observation interval τ.

With reference to Figure 9.8, consider two consecutive observation periods each consisting of N clock cycles. The first period, indicated with a, starts with the edge at time t_k and ends at time t_{k+N}. The second, indicated by b, starts at t_{k+N} and ends at t_{k+2N}. The average clock period across a, $\overline{T_a}$ can be calculated as the difference between the instants of the last and first edges, divided by the number of cycles:

$$\overline{T}_a = \frac{t_{k+N} - t_k}{N}. \tag{9.43}$$

Expressing t_k as the sum of the ideal clock plus absolute jitter $t_k = kT + \mathbf{a}_k$, and replacing it in the expression above, we find that:

$$\overline{T}_a = \frac{(k+N)T + \mathbf{a}_{k+N} - kT - \mathbf{a}_k}{N} = T + \frac{\mathbf{a}_{k+N} - \mathbf{a}_k}{N} \tag{9.44}$$

so that the normalized frequency deviation can be expressed in terms of the absolute jitter and of the N-period jitter as:

$$\overline{y}_a = -\frac{\mathbf{a}_{k+N} - \mathbf{a}_k}{NT} = -\frac{\mathbf{p}_k(N)}{NT}. \tag{9.45}$$

Following the same steps, the normalized frequency deviation across the observation period b, can be expressed as:

$$\overline{y}_b = -\frac{\mathbf{a}_{k+2N} - \mathbf{a}_{k+N}}{NT} = -\frac{\mathbf{p}_{k+N}(N)}{NT}. \tag{9.46}$$

Finally, the quantity calculated in Equation 9.39 can be written as:

$$\Delta y = \frac{\overline{y}_b - \overline{y}_a}{\sqrt{2}} = \frac{\mathbf{p}_k(N) - \mathbf{p}_{k+N}(N)}{\sqrt{2}NT} \tag{9.47}$$

and the ADEV:

$$\sigma_{\Delta y}(NT) = \frac{1}{NT}\sqrt{\frac{\mathrm{E}\left[(\mathbf{p}_k(N) - \mathbf{p}_{k+N}(N))^2\right]}{2}}. \tag{9.48}$$

From these expressions it can be seen that the ADEV over $\tau = NT$ is essentially a measure of the variability of the N-period jitter over two consecutive N-cycle observation periods, normalized to the observation interval.

9.3.4 Relation Between Allan Deviation and Phase Noise

In this section the relation between ADEV and phase noise will be derived by applying the general method outlined in Section 9.1. Starting from Equation 9.47, Δy can be written as a function of the absolute jitter as:

$$\Delta y(NT) = \frac{-(\mathbf{a}_k - 2\mathbf{a}_{k+N} + \mathbf{a}_{k+2N})}{\sqrt{2}NT}. \tag{9.49}$$

The expression above can be reformulated in terms of a discrete time signal $\mathbf{a}_k$ filtered by a FIR filter with appropriate coefficients:

$$\Delta y(NT) = \frac{-\mathbf{a}_k}{\sqrt{2}NT}(1 - 2z^{-N} + z^{-2N}) = \frac{-\mathbf{a}_k}{\sqrt{2}NT}(1 - z^{-N})^2 \tag{9.50}$$

where z^{-1} is the unity delay operator, which in this case represents a delay of one clock period. Using the relation between absolute jitter and excess phase $\mathbf{a}_k = -\varphi_k/\omega_0$ (see Equation 3.5) and basic theory of linear systems (see Appendix A.2.10), it is easy to derive the relation between the PSD of Δy and the PSD of the excess phase, which is by definition equal to the phase noise, as:

$$S_{\Delta y}(f) = \frac{|1 - z^{-N}|^4}{2\omega_0^2 \tau^2} \mathcal{L}(f) \tag{9.51}$$

where NT has been replaced by the more commonly used τ. By replacing $z^{-1} = e^{-j2\pi fT}$ in Equation 9.51, after some easy mathematical steps, finally:

$$S_{\Delta y}(f) = \frac{2 \sin^4(\pi f \tau)}{(\pi f_0 \tau)^2} \mathcal{L}(f) \tag{9.52}$$

and the relationship between ADEV and phase noise can then be written as:

$$\sigma_{\Delta y}(\tau) = \frac{2}{\pi f_0 \tau} \sqrt{\int_0^{+\infty} \sin^4(\pi f \tau) \mathcal{L}(f) \, df}. \tag{9.53}$$

From the previous equation it can be seen that different phase noise profiles lead to different behaviors of the ADEV versus τ. Therefore, in metrology, the ADEV has been used not only as a measure of the frequency stability of clock signals, but also as tool for characterizing the behavior in the frequency domain of the excess phase noise affecting the oscillator. Generic phase noise profiles can by represented by the power law $\mathcal{L}(f) = f^\alpha$, where α is an integer coefficient equal to or smaller than zero. Figure 9.9 gives an overview of the relation between the phase noise power law profiles and the different behaviors of the ADEV versus τ. Each of the profiles has a different "signature," with the exception of white phase modulation and flicker phase modulation, represented by $\alpha = 0$ and $\alpha = -1$, respectively, which lead to almost identical ADEV profiles. For this reason, in the early 1980s, a variant of the ADEV, the Modified Allan Deviation (MDEV), was defined, which can distinguish between these two sources of noise. This will be the topic of the next section.

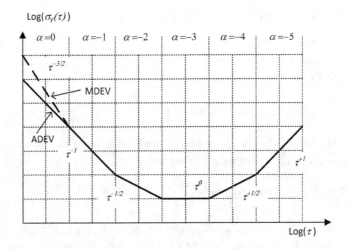

Figure 9.9 Relation between ADEV and phase noise power law.

9.3.5   Modified Allan Deviation (MDEV) and Time Deviation (TDEV)

The difference between the ADEV and the MDEV is that the MDEV computation involves an additional averaging of the $\bar{y}_i$.

With reference to Figure 9.10, in computing the MDEV, the observation interval τ is subdivided into n intervals τ_0. The $\bar{y}_i$ are then calculated over a window of length τ, shifted by multiples of τ_0. In this way, instead of having only one value of $\bar{y}_i$ within τ, n values are generated. These values are then used to calculate their first difference averaging over all n values. The first difference is then normalized by $\sqrt{2}$ as in the ADEV:

$$\overline{\Delta}y_i := \frac{1}{\sqrt{2}n} \sum_{i=0}^{n-1} (\bar{y}_{i+n} - \bar{y}_i). \tag{9.54}$$

The MDEV is calculated as the standard deviation of $\overline{\Delta}y$ and is often indicated in the literature as mod $\sigma_y(\tau)$.

From Equation 9.45, it is apparent that the MDEV is a measure of the variability of the N-period jitter averaged over n sliding windows within the observation interval $\tau = NT$, and that a relation very similar to Equation 9.48 holds:

$$\sigma_{\overline{\Delta}y}(NT) = \frac{1}{NT} \sqrt{\frac{E\left[\left[\frac{1}{n}\sum_{i=0}^{n-1} (\mathbf{p}_{k+iM}(N) - \mathbf{p}_{k+iM+N}(N))\right]^2\right]}{2}} \tag{9.55}$$

where $M = N\tau_0/\tau$.

As can be seen from Equations 9.48 and 9.55, both ADEV and MDEV are quantities carrying the information of the accuracy of the clocks in the time domain normalized to the observation interval $\tau = NT$. In the clock distribution for telecommunication networks, the goal is not only stability in frequency, but also synchronicity of the several clock signals distributed over different nodes of the network. For this reason another quantity, based on MDEV, was defined. This quantity is called time deviation (TDEV), and is simply defined as:

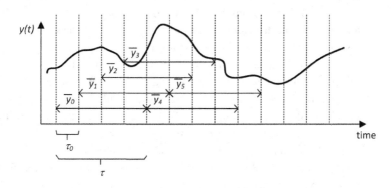

Figure 9.10 Computation of MDEV for the case $n = 4$.

$$\text{TDEV}(\tau) = \frac{\tau}{\sqrt{3}} \text{ MDEV}(\tau). \tag{9.56}$$

By denormalizing MDEV with respect to τ, the TDEV is a direct measure of the timing accuracy, rather than of the frequency accuracy.

9.4   An Overview of Flicker Noise

Phenomena which display a PSD of the form $S(f) \propto 1/f^\beta$, with $\beta \geq 1$ present us with some challenges, the foremost important one being how to consider the divergent integral of the PSD down to zero frequency. It is common knowledge that the integral of the PSD over frequency gives the power of the signal, and for this class of signals the integration seems to imply infinite power. This conclusion is obviously wrong, since no finite system can deliver infinite power.

A particular case of this phenomenon is the case of flicker noise, where the PSD assumes the form $S(f) \propto 1/f$. This kind of dependency of the PSD on the frequency has been observed not only in the noise in electronic systems, but also in many other areas, such as, among others, the average seasonal temperature, variation in economic data, and the loudness distribution over frequency in a musical piece. The phenomenon of $1/f$ noise is therefore very widespread, and by no means unique to the field of electronics. Theories have been developed to explain the phenomenon in particular application cases, but, to the author's knowledge, there's no generally accepted explanation of the basic fundamental mechanism at its origin.

Before delving into more elaborated considerations, let's face the problem of the apparent infinite power of flicker noise processes from a practical perspective. Following the consideration of Flinn [111], even assuming that the integral of the $1/f$ PSD over frequency delivers the power (not an obvious conclusion, as we will see later), we note that:

$$\int_{f_1}^{f_2} \frac{1}{f} df = \log\left(\frac{f_2}{f_1}\right) \tag{9.57}$$

so the amount of energy in a given range is only dependent on the ratio of the extreme frequencies. This means that the energy content in any frequency decade, no matter where in the frequency spectrum, is the same as the energy content between, for instance, 1Hz and 10Hz. Now, from a practical perspective, how many decades of the frequency spectrum should we consider in the worst case? Taking the assumed lifespan of the known universe (according to the Big-Bang theory, 14 billion years), roughly 10^{17}s, as the longest practical time, and the time it takes light to travel past the classical electron radius, 10^{-23}s, as the shortest time, we see that the number of decades covered by these two extremes is about 40. So, even in the worst case, the total energy content in a flicker noise phenomenon can be at most 40 times larger than its content in the 1Hz to 10Hz range. As an example, the flicker noise of a n-channel MOS device can be modeled as a voltage noise source at the gate with PSD (see [112]):

$$S_{vf}(f) = \frac{K}{C_{ox}^2 WLf} \qquad (9.58)$$

where, for a modern CMOS technology, K is a constant in the range of 10^{-26} C^2/m^2 and C_{ox} is of the order of 20 fF/um^2. For small device dimensions $W = 1um$ and $L = 40nm$, the integrated flicker noise voltage over one decade is of the order of 25uV RMS. Even considering the maximum practical frequency range of 40 decades, the total voltage noise would sum up to 1mV RMS, definitely a relevant value for some applications, but not at all threatening.

Despite this reassuring consideration, there have been continuous attempts in the past to find an indication of a flattening out of the flicker noise spectrum for very low frequencies. If this were the case, the integral of the PSD would be a finite number, solving the issue of the infinite power. As of today, in almost all cases investigated, no evidence has been found of any change of shape of the spectrum for very low frequencies. The authors in [113] investigated the $1/f$ noise in MOSFET down to $10^{-6.3}$Hz, and the $1/f$ trend was still on going. The only exceptions to this rule have been found in the resistance fluctuations of thin-films made of tin at the temperature of the superconducting transition and in the voltage fluctuations across nerve membranes.

Therefore, for all practical aspects, we have to accept that the $1/f$ noise extends far below the frequency range that can be measured with reasonable accuracy.

One of the most popular theories about the origin of flicker noise in electronic devices involves the fluctuation of current value due to the trapping and releasing of charges with different time constants. This theory, first proposed by Schottky [114] and subsequently refined by Bernamont [115], can be summarized as follows. In a conducting medium, charge packets are released at random instants distributed over time as Poisson points with parameter λ (average points per unit of time). Each charge packet generates a current which has an exponential decaying behavior of the form:

$$h(t) = h_0 e^{-\alpha t}. \qquad (9.59)$$

The total current, given by the superposition of the packets, fluctuates around the average value equal to $\lambda h_0/\alpha$. This process can be modeled as a impulse generator, triggered by a Poisson process, followed by a linear filter with impulse response $h(t)$, as shown in Figure 9.11. It can be shown (see, e.g., [1]) that the PSD of the Poisson impulse process z is given by:

$$S_z(\omega) = 2\pi \lambda^2 \delta(\omega) + \lambda \qquad (9.60)$$

and, using linear filter theory, the PSD of the output current process can be expressed as:

$$S_i(\omega) = 2\pi \lambda^2 H^2(0)\delta(\omega) + \lambda |H(\omega)|^2 \qquad (9.61)$$

where $H(\omega)$ is the Fourier transform of $h(t)$. In this expression, the first term represent the DC term, the average current value, while the second is the associated noise. In the particular case of an exponential decay, as in Equation 9.59:

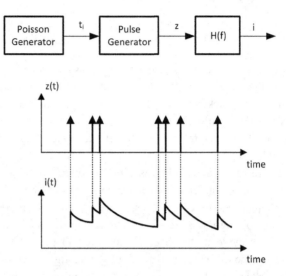

Figure 9.11 Model of a flicker noise current generator based on the theory of charge trapping and releasing.

$$H(\omega) = \frac{h_0}{\alpha + j\omega} \tag{9.62}$$

so that the PSD of the noise current is:

$$S(\omega) = \frac{\lambda h_0^2}{\alpha^2 + \omega^2} \tag{9.63}$$

showing a flat frequency behavior at low frequencies, and decaying as $1/f^2$ for $f > \alpha/(2\pi)$. If one postulates the existence of many trapping sites in the conducting medium, with time constants α uniformly distributed between two extremes α_1 and α_2, the compound PSD is given by the linear superposition of the PSD of the single processes:

$$S_n(\omega) = \frac{1}{\alpha_2 - \alpha_1} \int_{\alpha_1}^{\alpha_2} \frac{\lambda h_0^2}{\alpha^2 + \omega^2} d\alpha = \frac{\lambda h_0^2}{\omega(\alpha_2 - \alpha_1)} \left[\arctan\left(\frac{\alpha_2}{\omega}\right) - \arctan\left(\frac{\alpha_1}{\omega}\right) \right] \tag{9.64}$$

where in this case λ is the compound impulse rate. This expression can be approximated as:

$$S_n(\omega) = \begin{cases} \dfrac{\lambda h_0^2}{\alpha_1 \alpha_2} & \text{if } \omega \ll \alpha_1 \\[2ex] \dfrac{\pi \lambda h_0^2}{2\omega(\alpha_1 - \alpha_2)} & \text{if } \alpha_1 \ll \omega \ll \alpha_2 \\[2ex] \dfrac{\lambda h_0^2}{\omega^2} & \text{if } \omega \gg \alpha_2. \end{cases} \tag{9.65}$$

It can be seen that for frequencies between $\alpha_1/(2\pi)$ and $\alpha_2/(2\pi)$ the PSD follows a $1/f$ profile. It has to be noted, though, that for this model to explain the $1/f$ behavior below the lowest frequency currently measurable, one would require to postulate the existence of trapping sites with extremely long relaxation time $1/\alpha$, longer than what can be measured with modern techniques and, potentially, infinitely long. This merely exchanges one infinity with another, and therefore does not seem to be a satisfactory solution to the issue.

In [116] Mandelbrot made the key observation that a process whose PSD has a $1/f^\beta$ behavior with $\beta \geq 1$ cannot be stationary. Indeed, we know that the $1/f^2$ spectrum is typical of random walks, or Brownian motion, where the variance of the process increases linearly with time, and a time-dependent variance is an obvious sign of a non-stationary process. Measurements made on the variance of signal affected by flicker noise showed that the variance increases logarithmically with time (see, e.g., [117]), so that also flicker noise should be considered to be non-stationary. Since the process is not stationary, the Wiener–Khinchin theorem, stating that the PSD is the Fourier transform of the autocorrelation, does not hold. The consequence is that the integral of the PSD cannot be assumed to represent the power of the signal, which is, in the first place, given by the value of the autocorrelation for a time delay equal to zero.

Nevertheless, common practice in the industry is to calculate the power of the flicker noise by integrating the corresponding PSD, as if the process were stationary.

Another aspect often encountered is that repeated experimental measuring of the PSD of a flicker noise process, apparently under identical conditions, reveals that the slope of the PSD is pretty constant, but the magnitude can often change by orders of magnitude between one measurement and the next.

In [118] Keshner made an attempt to explain both aspects, by deriving an expression for the autocorrelation of a specific flicker noise process. The system investigated is an infinite-length distributed RC line, driven by a white noise current source. It is known that the characteristic impedance Z_0 of a generic transmission line is given by:

$$Z_0 = \sqrt{\frac{R + j\omega L}{G + j\omega C}} \tag{9.66}$$

where R, L, G, and C are the resistance, inductance, conductance loss, and capacitance per unit length. If we assume $L = 0$ and $G = 0$, then the characteristic impedance assumes the form:

$$Z_0 \propto \frac{1}{\sqrt{jf}}. \tag{9.67}$$

By applying traditional linear system theory, if this line is driven by a white noise current source with PSD $S_i(f)$, the output voltage PSD $S_v(f)$ will be:

$$S_v(f) = S_i(f) \cdot |Z_0|^2 \propto \frac{1}{f} \tag{9.68}$$

so that the PSD of the voltage at the line input follows perfectly a $1/f$ profile down to zero frequency. Instead of following this approach, in [118] the author proceeds to calculate the autocorrelation function of the voltage $v(t)$ directly in the time domain, under the assumption that the current source is a impulse shot noise process with zero mean. The response of the line to a current impulse at time t_0 is simply the inverse Fourier transform of Equation 9.67, which is:

$$v(t) \propto \frac{1}{\sqrt{t - t_0}} \tag{9.69}$$

for $t > t_0$ and zero otherwise. Assuming that the time t_0 is a uniformly distributed random variable in the interval $[0, T]$, $v(t)$ becomes a random process, and its autocorrelation function, defined as $R_v(t_1, t_2) := \mathrm{E}\,[v(t_1)v(t_2)]$ can be calculated as:

$$R_v(t_1, t_2) = \int_{-\infty}^{+\infty} \frac{1}{\sqrt{t_1 - t_0}} \frac{1}{\sqrt{t_2 - t_0}} \cdot (\text{PDF of } t_0) dt_0$$

$$\propto \int_0^{t_1} \frac{1}{\sqrt{t_1 - t_0}} \frac{1}{\sqrt{t_2 - t_0}} dt_0 \tag{9.70}$$

where it has been assumed that $t_1 < t_2 < T$. This integral can be evaluated with the help of integral tables (see, e.g., [119]) and is equal to:

$$\log\left(\frac{\sqrt{t_2} + \sqrt{t_1}}{\sqrt{t_2} - \sqrt{t_1}}\right). \tag{9.71}$$

If the line has been connected to the current source long before the voltage at its terminals is measured, then $0 \ll t_1 < t_2$ and $t_2/t_1 \approx 1$ so that finally the autocorrelation function is:

$$R_v(t_1, t_2) \propto \log\left(\frac{4t_2}{t_2 - t_1}\right) = \left[\log(4t_2) - \log(|\tau|)\right] \tag{9.72}$$

where $\tau := t_2 - t_1$ according to the widespread usage. Based on this result, the conclusion of the author in [118] is that the autocorrelation function of this process is "almost" stationary. The nonstationary behavior is entirely contained in the added term that depends only on t_2. By taking the Fourier transform of 9.72 with respect to τ, the PSD includes a $1/|f|$ term coming from the $\log(|\tau|)$ plus an impulse at frequency zero, whose amplitude depends on the logarithm of the time elapsed since the system is active t_2. The author argues that a careful observer would be able to obtain repetitive measurements of the PSD of the system, by allowing in each measurement attempt a roughly constant, sufficiently long time to pass from the moment the system is activated to the moment the measurement is taken. According to the author, the almost stationary autocorrelation function of the form in Equation 9.72 would also explain some experimental observations where the value of the slope of the PSD varies slightly but the value of the magnitude sometimes varies enormously.

For the interested reader, [120] explains some of the concepts outlined in this section, presenting them in a fashion more familiar to electrical engineers.

Bringing this conversation back to the framework of jitter and phase noise, there is one key observation to be made: there is no contradiction between a phase noise spectrum with infinite power and a voltage or current spectrum with finite power, and the former is the only one relevant when jitter or phase noise is considered. The concept of a phase noise with infinite power should not horrify the reader. There is no energy in the phase of a signal; all the energy is contained in the amplitude.[2] Infinite phase noise power just

---

[2] Recall also another similar apparent paradox: the phase velocity of an electromagnetic wave can be larger than the speed of light. Another similar consideration apply to the point of breaking of a sea wave on the

means that, if one waits long enough, the phase of a real signal will drift further and further from the ideal phase.

9.5   Lorentian Spectrum of Oscillator Phase Noise

In Section 4.3, it was shown how the phase noise of a free-running oscillator follows a $1/f^2$ profile. Similar to the subject of the previous section, this behavior poses the fundamental problem that the PSD diverges to infinity for very low values of the frequency offset f. The methods used in that section, however, included some approximations, which are acceptable for large frequency offsets but do not hold for very small ones. In this section, a more precise analysis shows that the spectrum of a free-running oscillator does indeed have a *Lorentian* profile: it is flat for frequencies below a given corner frequency and rolls off as $1/f^2$ above it. The derivation follows the original work of Demir [6], including a few heuristic arguments in order to keep it shorter. The reader is referred to the original paper for an exact mathematical treatise. The interested reader can also refer to [121] for a treatment using concepts probably more familiar to designers.

Let's consider a signal $x(t)$, periodic in $2\pi/\omega_0$. The signal can be expanded in Fourier series as:

$$x(t) = \sum_{n=-\infty}^{+\infty} X_n e^{jn\omega_0 t}. \qquad (9.73)$$

Assume that the signal is subject to excess phase noise, but not to amplitude noise. This can be modeled by adding a time-dependent noise component $\alpha(t)$ to the ideal time axis t, so that the noisy signal can be written as $x(t+\alpha(t))$. As an example, a sinusoidal signal with added excess phase $\phi(t)$ can be written as $\sin(\omega_0 t + \phi(t)) = \sin(\omega_0(t+\phi(t)/\omega_0))$, so that in this case $\alpha(t) = \phi(t)/\omega_0$. The autocorrelation of the noisy signal is by definition:

$$R_x(t,\tau) := E\left[x(t+\alpha(t)) \cdot x^*(t+\tau+\alpha(t+\tau))\right] \qquad (9.74)$$

and is in general, not stationary, since it depends on time t. By replacing Equation 9.73 in Equation 9.74, the autocorrelation assumes the form:

$$R_x(t,\tau) = \sum_{n,m} X_n X_m^* e^{j(n-m)\omega_0 t} e^{-jm\omega_0\tau} E\left[e^{j\omega_0[n\alpha(t)-m\alpha(t+\tau)]}\right]. \qquad (9.75)$$

To evaluate this result further, it is necessary to find expressions for the expectation term $E\left[e^{j\omega_0[n\alpha(t)-m\alpha(t+\tau)]}\right]$. The key observation is that the excess phase of a free-running oscillator is produced by the steady accumulation of uncorrelated circuit noise over cycles. Thus $\alpha(t)$ has the statistical characteristic of a Gaussian random walk process; or, more precisely a Wiener process if the added noise is white. Having said that, it is straightforward to understand that, for the case $n \neq m$, the two processes $n\alpha(t)$ and $m\alpha(t+\tau)$ diverge from each other. Their difference grows indefinitely over time, and

shore. If the angle between the wave and the line of the shore (assumed both to be straight) is extremely small, the point of breaking can travel along the shore very fast, even faster than light (at least in a thought experiment). But again, there is no contradiction, since the point of breaking does not transport any energy.

applied to the argument of a sinus or cosinus functions (as in the complex exponential under consideration) results in a number with average, i.e., expected value, equal to zero. This heuristic derivation can be proven mathematically as in [6]. For $n = m$ the expectation term $\mathrm{E}\left[e^{j\omega_0[n\alpha(t)-m\alpha(t+\tau)]}\right]$ can be evaluated by considering the random process $z(\tau)$:

$$z(\tau) := n\left[\alpha(t) - \alpha(t+\tau)\right]. \tag{9.76}$$

This process is obtained by taking the difference of the process $\alpha(t)$ at two time instants separated by τ. Since $\alpha(t)$ is a random walk process, it can be seen that $\alpha(t) - \alpha(t+\tau)$ is non stationary Gaussian process with zero mean and variance linearly increasing with time as $c\tau$ [1]. Thus $z(\tau)$ is a zero-mean Gaussian random variable with variance equal to $n^2c\tau$. By using Equation 9.76 the expectation term under consideration can be written as $\mathrm{E}\left[e^{j\omega_0 z(\tau)}\right]$. This turns out to be the *characteristic function* of a Gaussian random variable evaluated at ω_0, and it can be proved to be (see, e.g., [1]):

$$\mathrm{E}\left[e^{j\omega_0 z(\tau)}\right] = e^{-\frac{1}{2}\omega_0^2 n^2 c|\tau|}. \tag{9.77}$$

Summarizing the previous considerations, the time average over t of $\mathrm{E}\left[e^{j\omega_0[n\alpha(t)-m\alpha(t+\tau)]}\right]$ is equal to:

$$\begin{cases} 0 & \text{if } n \neq m \\ e^{-\frac{1}{2}\omega_0^2 n^2 c|\tau|} & \text{if } n = m. \end{cases} \tag{9.78}$$

By replacing the previous expression into Equation 9.75, the autocorrelation averaged over time results in:

$$\overline{R_x}(\tau) = \sum_n |X_n|^2 e^{-jn\omega_0\tau} e^{-\frac{1}{2}\omega_0^2 n^2 c|\tau|} \tag{9.79}$$

and, by taking the Fourier transform of the autocorrelation, the spectrum of the signal $x(t+\alpha(t))$ can be expressed as:

$$S_x(f) = \sum_n |X_n|^2 \frac{\omega_0^2 n^2 c}{\frac{1}{4}\omega_0^4 n^4 c^2 + (\omega + n\omega_0)^2}. \tag{9.80}$$

It can be seen that for frequencies far away from the harmonics ($\omega \gg n\omega_0 + \omega_0^2 n^2 c/2$) the spectrum as a $1/f^2$ profile, while it flattens out at offset frequencies below $\omega_0^2 n^2 c/2$. It is also interesting to note how the integral in Equation 9.80 around each harmonic is equal to the power of the harmonic itself $|X_n|^2$. This means that the excess phase noise does not change the energy of the signal; it just spreads its energy over the frequency spectrum, blurring the very sharp frequency impulse of an ideal noiseless oscillator. Also observe how the expression above is in agreement with Equation 3.20, stating the relation between the phase noise of the fundamental and that of the harmonics, obtained with other considerations.

10 Numerical Methods

The most common approach to the investigation of the behavior of electronic systems in the presence of jitter is to use linear modeling in the frequency domain and phase noise profiles. However, especially in the presence of complex or non-linear systems, it is often beneficial to perform transient simulations and evaluate the results in the time domain. To use this approach, it is necessary to provide numerical methods to, first, generate clock signals in the time domain that have given phase noise profiles and, second, analyze the results of transient simulations and possibly convert them into phase noise plots. The goal of this chapter is to provide some of these techniques.

Most of the chapter is dedicated to algorithms and numerical recipes to generate jitter samples with several different phase noise profiles, and to compute jitter and phase noise given a vector of time instants, typically obtained from transient simulations. Following that, basic algorithms to perform tail fitting are presented. All algorithms are implemented and demonstrated using Matlab and are coded such that they can be easily ported to any other common programming language. The code is included in Appendix B.

10.1 Numeric Generation of Jitter with Given Phase Noise Profiles

The problem addressed in this section is how to generate a sequence of numbers t_k representing the instants of the edge transitions of a clock having nominal frequency f_0 and a desired phase noise profile $\mathcal{L}(f)$.

As a starting point, the sequence t_k can be decomposed into the sum $t_k = t_{id,k} + \mathbf{a}_k$ of a vector describing an ideal clock, $t_{id,k} = k/f_0$ with $k \geq 0$, and a vector $\mathbf{a}_k$ describing the jitter process. Note that the $\mathbf{a}_k$ corresponds to the absolute jitter, according to the definition given in Section 2.1.1. The problem translates then into the generation of the sequence $\mathbf{a}_k$ showing a given phase noise profile $\mathcal{L}(f)$.

Before proceeding it is important to recall that the PSD of the sequence $\mathbf{a}_k$ is independent from its probability density function (PDF). To make it clearer, imagine one realization of the process $\mathbf{a}_k$ being plotted on a graph, where the x-axis is the index k and the y-axis is the value (amplitude) of $\mathbf{a}_k$. The PSD describes the behavior of the sequence in the frequency domain and is thus connected to how the samples $\mathbf{a}_k$ develop along the x-axis. The PDF, on the other hand, is a histogram of the amplitude of the process and thus a description of how the samples are distributed along the y-axis. In principle, a white process (flat PSD) can have different PDF (Gaussian, uniform, triangular, etc.)

and, conversely, a process with a certain PDF can assume several different PSD shapes, depending on its behavior over time. Although Gaussian (Gaussian PDF) white (flat PSD) processes are very common, a Gaussian process is not necessarily white, and a white process is not necessarily Gaussian.

In the following sections, we will make use of the relationship between the PSD of jitter S_a and the phase noise $\mathcal{L}(f)$, derived in Section 3.1.2 and repeated here again for convenience:

$$S_a(f) = \frac{\mathcal{L}(f)}{\omega_0^2}. \tag{10.1}$$

10.1.1   Generation of Jitter Samples with Flat Phase Noise Profile

Assume the desired $\mathcal{L}(f)$ is flat over a frequency range and equal to $\mathcal{L}_0$ as shown in Figure 3.10. It follows from Equation 10.1 that the sequence a_k must also have a flat PSD. It is thus natural to use as a starting point a sequence a_k obtained from a Gaussian pseudo-random number generator with mean zero and variance σ_a^2. The question is now how to choose σ_a so that $\mathcal{L}(f) = \mathcal{L}_0$. Since a_k is a discrete time white noise process defined on time instants that are multiples of $1/f_0$, its variance is related to the PSD by:

$$S_a(f) = \frac{\sigma_a^2}{f_0}. \tag{10.2}$$

By equating this expression to Equation 10.1, we obtain:

$$\sigma_a = \frac{1}{2\pi} \sqrt{\frac{\mathcal{L}_0}{f_0}}. \tag{10.3}$$

For an example of Matlab code, see Section B.1.1.

10.1.2   Generation of Jitter Samples with Low-Pass, High-Pass, or Band-Pass Phase Noise Profiles

In most practical situations it is necessary to generate jitter samples with specific profiles, such as high-pass, band-pass, or low-pass. In the case of a simple PLL spectrum, such as the one shown in Figure 3.12, a low-pass profile is needed. These profiles can be generated quite easily by taking the jitter samples for a flat profile (as described in Section 10.1.1) and then filtering them using a digital filtering function, which is available in most mathematical packages, to shape the jitter samples accordingly. If the digital filter implements the transfer function $H(f)$, then the resulting phase noise will be $\mathcal{L}(f) = |H(f)|^2 \mathcal{L}_0$.

For an example of Matlab code, see Section B.1.2.

10.1.3   Generation of Jitter Samples with $1/f^2$ Phase Noise Profile

Free-running oscillators show phase noise profiles which have a $1/f^2$ or $1/f^3$ characteristic close to the carrier. In this section we will address the generation of $1/f^2$ profiles such as those shown in Figure 3.11. Since in these cases there is no flat region, the approach of Section 10.1.2 cannot be used directly. However, from basic linear system theory, it is known that a $1/f^2$ spectrum can be obtained by passing a flat spectrum through an ideal integrator. Exploiting this idea, first a Gaussian sequence of jitter samples l_k with variance σ_l^2 and flat spectrum is generated. This sequence is then passed through an accumulator to generate the sequence $\mathbf{a}$, where $\mathbf{a}_k = \mathbf{a}_{k-1} + l_k$. The main question is how to choose the value of σ_l so that the profile shown in Figure 3.11 is obtained.

From the basics of random process theory, we know that the PSD of the sequence l is $S_l(f) = \sigma_l^2/f_0$, and from basic linear theory (see also Section 9.1):

$$S_{\mathbf{a}}(f) = \frac{1}{\left|1 - e^{-j2\pi f/f_0}\right|^2} S_l(f). \tag{10.4}$$

Using the fact that the region of interest is where $f \ll f_0$, and that $\mathcal{L}(f) = S_{\mathbf{a}}(f)\omega_o^2$, the resulting phase noise profile can be expressed as:

$$\mathcal{L}(f) = \left(\frac{f_0}{2\pi f}\right)^2 \frac{\sigma_l^2}{f_0} \omega_0^2. \tag{10.5}$$

By equating this expression to the desired phase noise profile $\mathcal{L}(f) = \mathcal{L}_1 f_1^2/f^2$, as shown in Figure 3.11, and solving for σ_l, finally:

$$\sigma_l = \frac{f_1}{f_0}\sqrt{\frac{\mathcal{L}_1}{f_0}}. \tag{10.6}$$

For an example of Matlab code, see Section B.1.3.

10.1.4 Generation of Jitter Samples with $1/f$ and $1/f^3$ Phase Noise Profiles

Spectral densities with a $1/f$ profile are typical of flicker noise processes. In this section, we will first explain a method to generate a noise profile of the form $\mathcal{L}_1 f_1/f$. In a second step, by applying a further integration step, the $1/f$ PSD will be converted into a $1/f^3$ PSD.

In order to generate the $1/f$ profile we will use the following procedure. A random sequence l_k with a flat PSD and a variance σ_l will be passed through a bank of first-order low-pass filters, whose outputs will be summed to give the sequence $\mathbf{a}_k$. It is natural to think that if the DC gains and the corner frequencies of the low-pass filters are scaled such that a reduction of the gain by a factor α is accompanied by an increase of the corner frequency by a factor α^2, the transfer function of the complete bank will resemble a $1/f$ profile (see Figure 10.1). Indeed, it is known (see [122]) that, in order to obtain a pretty accurate approximation of a $1/f$ profile, it is sufficient to implement one single low-pass filter per decade of frequency, meaning that a factor $\alpha = \sqrt{10}$ is enough. This representation of $1/f$ noise was first introduced by Bernamont [115] and is based on the most widespread underlying principle of various models describing $1/f$ noise generation in several physical phenomena. A more detailed description of this approach applied to noise in oscillators can be found in [123].

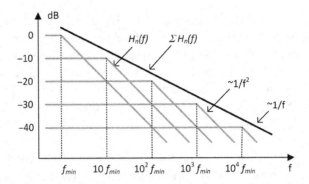

Figure 10.1 Composition of first-order low-pass filters to obtain a $1/f$ noise profile.

With reference to Figure 10.1, if the flicker noise profile is to be generated over a frequency range from f_{min} to f_{max}, the corner frequencies of the bank of filters should span at least the same interval. For simplicity, f_{max} is chosen to be an integer number N of decades away from f_{min} ($f_{max} = 10^N f_{min}$). The transfer function $H_n(f)$ of the n-th bank can be expressed as:

$$H_n(f) = \frac{10^{-n/2}}{1 + j\frac{f}{10^n f_{min}}}. \tag{10.7}$$

The PSD $S_{\mathbf{a}}(f)$ of the output process $\mathbf{a}_k$ can be expressed as:

$$S_{\mathbf{a}}(f) = \left| \sum_{n=0}^{n=N} H_n(f) \right|^2 S_l(f) \tag{10.8}$$

where $S_l(f) = \sigma_l^2/f_0$ is the PSD of the input process l_k.

In order to simplify this expression, consider the value of $\sum H_n(f)$ for frequencies in the proximity of the corner frequency of a given n_0-th bank, $10^{n_0} f_{min}$. Around this frequency, the major contributions to $\sum H_n(f)$ come from the n_0-th bank itself and from the two adjacent banks, so that:

$$\left| \sum_{n=0}^{n=N} H_n(10^{n_0} f_{min}) \right|^2 \approx \left| H_{n_0-1}(10^{n_0} f_{min}) + H_{n_0}(10^{n_0} f_{min}) + H_{n_0+1}(10^{n_0} f_{min}) \right|^2. \tag{10.9}$$

Replacing Equation 10.7 in the previous expression:

$$\left| \sum_{n=0}^{n=N} H_n(10^{n_0} f_{min}) \right|^2 \approx \frac{1}{10^{n_0}} \left| \frac{\sqrt{10}}{1 + j10} + \frac{1}{1 + j} + \frac{\sqrt{10}}{10 + j} \right|^2 = \frac{\gamma}{10^{n_0}} \tag{10.10}$$

with $\gamma = 1.426$. This means that $\left| \sum H_n(f) \right|^2$ is taking the value $\gamma/10^{n_0}$ at $10^{n_0} f_{min}$, and since the assumption is that it shows a $1/f$ profile, the following expression can be derived:

$$\left| \sum_{n=0}^{n=N} H_n(f) \right|^2 = \gamma \frac{f_{min}}{f}. \tag{10.11}$$

Replacing the expression above in Equation 10.8, and considering the general relationship between the PSD of jitter and phase noise (Equation 10.1), the phase noise spectrum of the resulting jitter process can be finally expressed as:

$$\mathcal{L}(f) = \frac{4\pi^2 f_0 \sigma_l^2 \gamma f_{min}}{f}. \tag{10.12}$$

It can be seen that, in order to obtain the desired $\mathcal{L}(f) = \mathcal{L}_1 f_1 / f$, the variance of the input noise process l_k has to be selected to be equal to:

$$\sigma_l^2 = \frac{\mathcal{L}_1 f_1}{4\pi^2 \gamma f_{min} f_0}. \tag{10.13}$$

For an example of Matlab code, see Section B.1.4.

In order to obtain a $1/f^3$ profile of the form $\mathcal{L}(f) = \mathcal{L}_1 f_1^3 / f^3$, it is sufficient to apply a further integration step to the sequence obtained above. As described in Section 10.1.3, the resulting PSD will show an additional factor $[f_0/(2\pi f)]^2$, and the final phase noise spectrum will be:

$$\mathcal{L}(f) = \frac{\sigma_l^2 \gamma f_{min} f_0^3}{f^3}. \tag{10.14}$$

In order to obtain the desired profile, the variance of the input process l_k must be selected to be equal to:

$$\sigma_l^2 = \frac{\mathcal{L}_1 f_1^3}{\gamma f_{min} f_0^3}. \tag{10.15}$$

For an example of Matlab code, see Section B.1.5.

10.1.5 Generation of Jitter Samples with More Complex Phase Noise Profiles

The basic profiles described in the previous sections can be easily combined to form more complex phase noise profiles. Indeed, if independent jitter processes are generated using the approaches above starting from uncorrelated random number sources and then summed together, the PSD of the sum will be equal to the sum of the PSD of the single processes. In this way, many different complex PSD shapes can be easily implemented.

As an example, the code in Section B.1.6 describes the generation of a phase noise profile featuring a $1/f^3$ region, a $1/f^2$ region, a PLL-like spectrum and, finally, a flat noise floor.

The resulting phase noise profile, obtained using the techniques explained in Section 10.2, is shown in Figure 10.2.

10.2 Computation of Jitter from Vector of Time Instants

The problem addressed here is the extraction of jitter information starting from an array of numbers, representing the edge instants of a jittered clock, which might be the result of a simulation or come from measurements. Assume an array t_k with $k = 0 \ldots n - 1$ representing the time instants of the rising edges of a jittered clock signal; then the

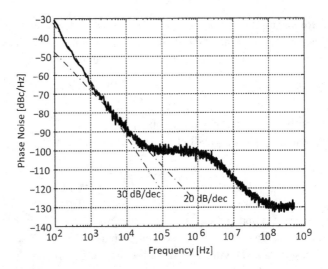

Figure 10.2 Example of a complex phase noise profile.

absolute jitter is defined as the time displacement of the k-th rising edge t_k with respect to the edge (kT) of an ideal clock having the same period T and a possible timing offset t_{OS}:

$$\mathbf{a}_k := t_k - kT - t_{OS}. \tag{10.16}$$

The problem now translates into finding appropriate expressions for T and t_{OS}. Since we have two unknowns, we need two equations to arrive at a unique solution. The two criteria we will use to write the equations are to require that, first, the absolute jitter has zero mean, and, second, the absolute jitter has the least variance. The first criterion translates into the following condition:

$$\sum_{k=0}^{n-1} \mathbf{a}_k = \sum_{k=0}^{n-1} (t_k - kT - t_{OS}) = 0 \tag{10.17}$$

which gives the following expression for t_{OS} as a function of T:

$$t_{OS} = \frac{1}{n} \sum_{k=0}^{n-1} (t_k - kT). \tag{10.18}$$

We now need to find T. The second criterion is equivalent to minimizing the following expression:

$$\sum_{k=0}^{n-1} \mathbf{a}_k^2 = \sum_{k=0}^{n-1} (t_k - kT - t_{OS})^2. \tag{10.19}$$

Note that, in writing this expression, we have already assumed that the average jitter is zero, otherwise we would have had to include the jitter average in the computation of the variance. Using Equation 10.18, this condition dictated by the second criterion can be expressed as:

$$\frac{\partial}{\partial T} \sum_{k=0}^{n-1} \mathbf{a}_k^2 = 2 \sum_{k=0}^{n-1} \mathbf{a}_k \frac{\partial \mathbf{a}_k}{\partial T} = 0. \tag{10.20}$$

From Equations 10.16 and 10.18, we have $\partial \mathbf{a}_k / \partial T = (n-1)/2 - k$. Since we assumed that $\mathbf{a}_k$ has zero mean, we can rewrite the previous equation as:

$$\sum_{k=0}^{n-1} k \mathbf{a}_k = 0. \tag{10.21}$$

Using Equations 10.16 and 10.18, after some manipulation, the following expression for T is obtained:

$$T = \frac{12}{n(n^2 - 1)} \sum_{k=0}^{n-1} \left(k - \frac{n-1}{2} \right) t_k. \tag{10.22}$$

On a first guess, it may appear that T should be equal to the average period given by the sequence t_k, which is $(t_{n-1} - t_0)/(n-1)$. The reader can easily confirm that this is indeed true for $n = 2$ and $n = 3$. For $n > 3$, however, Equation 10.22 shows that the correct T depends on the specific t_k distribution between t_0 and t_{n-1}, and may be larger or smaller than the average period. We demonstrate this with an example.

EXAMPLE 18   Assume $t_0 = 0$ and $t_4 = 4\text{ns}$. The average period is therefore 1ns irrespective of the values of t_1, t_2, and t_3. We now use Equation 10.22 to calculate T for the following three cases: (a) $t_1 = 1\text{ns}$, $t_2 = 2\text{ns}$, $t_3 = 3\text{ns}$, (b) $t_1 = 1.8\text{ns}$, $t_2 = 2\text{ns}$, $t_3 = 2.2\text{ns}$, (c) $t_1 = 0.5\text{ns}$, $t_2 = 2\text{ns}$, $t_3 = 3.5\text{ns}$. The T for cases (a), (b), and (c) are 1ns, 0.84ns, and 1.1ns, respectively.

The extraction of the ideal clock from the jittered data can also be approached in a different way. The time of the transitions of the clock can be plotted on a graph, as depicted in Figure 10.3, in which the value on the x-axis is the index k of the observed rising edge, and the y-axis is the time value at which the transition occurs. If the clock were ideal, the t_k values would all lie on a straight line with a slope equal to the period. In reality, because of jitter, the values deviate in a random manner. The goal is to find a

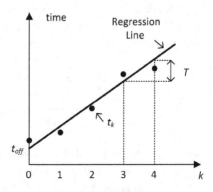

Figure 10.3 Time of the transitions vs. transition index.

straight line with the minimum mean square distance from the observed transition values. In statistics, this line is called the *best fit line*, *regression line*, or *least squares line*. In general, the least squares line associated with n points (x_k, y_k), with $k = 0 \ldots n - 1$, has the form $y = \beta_1 x + \beta_0$; see for instance [124], where:

$$\beta_1 = \frac{n \sum x_k y_k - \sum x_k \sum y_k}{n \sum x_k^2 - (\sum x_k)^2} \tag{10.23}$$

$$\beta_0 = \frac{\sum y_k - \beta_1 \sum x_k}{n}. \tag{10.24}$$

Setting $\beta_1 = T$, $\beta_0 = t_{OS}$, $y_k = t_k$, and $x_k = k$, the formulas for the slope and intercept of the regression line are identical to Equations 10.18 and 10.22. The two approaches are identical.

10.3   Computation of Phase Noise Plot from Jitter Samples

Once the jitter has been extracted from the data, the computation of the phase noise is fairly straightforward. Applying Equation 3.5, the jitter samples are first converted into excess phase samples and then a PSD operation is applied. Most mathematical packages have built-in functions to compute the PSD of arrays of data. Alternatively, the PSD can be calculated via FFT using standard approaches, as can be found, for instance, in [72].

10.4   Algorithms for Tail Fitting

In this section, the topic of tail fitting, already introduced in Section 2.2.8, is analyzed further, focusing on the basic mathematical methods involved. Specifically, we introduce and elaborate the concepts of *Q-scale* and *normalized Q-scale* transforms, which have proven to be very useful and are at the basis of the algorithms for tail fitting currently implemented in some specialized measurement equipment.

It has been recognized that the tails of the distribution resulting from the convolution of a Gaussian PDF with a bounded PDF, have a linear behavior if a particular transformation is applied. In this approach, the problem of Gaussian fitting can be reduced to a standard problem of linear fitting, and the mean and sigma values of the best Gaussian fit can be extracted very easily from the parameters of the straight line fitting the linearized distribution.

As starting point let's consider the equation describing the PDF of a Gaussian distribution:

$$n(x) = \frac{1}{\sigma \sqrt{2\pi}} \exp\left[-\frac{(x - \mu)^2}{2\sigma^2} \right]. \tag{10.25}$$

Note that in this section we will use the dimensionless variable x to denote the x-axis. While in the context of this book x is meant to be jitter, the methodology explained here is general and can be applied to many other fields. From the equation above, the corresponding Cumulative Distribution Function (CDF) can be computed as:

$$N(x) = \int_{-\infty}^{x} n(t)\, dt = \frac{1}{\sigma\sqrt{2\pi}} \int_{-\infty}^{x} \exp\left[-\frac{(t-\mu)^2}{2\sigma^2}\right] dt. \qquad (10.26)$$

With a simple change of variable $v = (t - \mu)/(\sqrt{2}\sigma)$ the above expression can be rewritten as:

$$N(x) = \frac{1}{\sqrt{\pi}} \int_{-\infty}^{\frac{x-\mu}{\sigma\sqrt{2}}} \exp(-v^2)\, dv \qquad (10.27)$$

and finally, using the complementary error function erfc already introduced in Section 9.2, Equation 9.16:

$$N(x) = \frac{1}{2}\text{erfc}\left(-\frac{x-\mu}{\sigma\sqrt{2}}\right). \qquad (10.28)$$

If we invert the formula above the following fundamental relationship is found:

$$-\sqrt{2}\,\text{erfc}^{-1}\,[2N(x)] = \frac{x-\mu}{\sigma}. \qquad (10.29)$$

At this point, let's define the Q-scale transform of a particular CDF $\Phi(x)$ as:

$$Q(x) := -\sqrt{2}\,\text{erfc}^{-1}\,[2\Phi(x)]. \qquad (10.30)$$

From Equation 10.29 it is clear that if the Q-scale transform is applied to the CDF of a Gaussian distribution, the result is a straight line, with slope $1/\sigma$ and intercepting the x-axis at μ (see Figure 10.4 for an example).

Supported by this insight, tail fitting can be performed by following some easy steps. First the CDF of the left and right tails are calculated based on the PDF or histogram of the measured data. In this sense note that the *left* CDF is the usual CDF (integral of the PDF starting from $-\infty$ up to a given x-axis value), while the *right* CDF is defined as

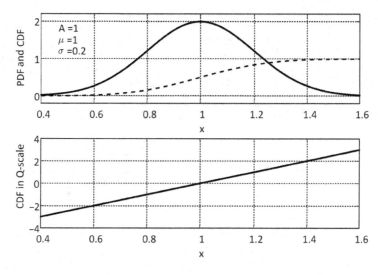

Figure 10.4 Linearization of the CDF of a Gaussian distribution, by use of the Q-scale transformation. Top: PDF and CDF of a Gaussian distribution with $\mu = 1$, $\sigma = 0.2$. Bottom: CDF plot obtained by applying the Q-scale transformation 10.30.

the integral of the PDF starting from a given x-axis value to $+\infty$. In a second step, the Q-scale transform in Equation 10.30 is applied to the left and right CDF. Third, a best linear fit (e.g., by using the LMS criteria) is applied to the linearized CDF and, finally, the values of the mean and sigma for the left and right tails are extracted from the fitted lines, based on Equation 10.29.

Figures 10.5 and 10.6 show the result of the procedure outlined above, applied to two cases. The first case is obtained by the convolution of a Gaussian curve having mean $\mu = 0$ and $\sigma = 0.4$ with a dual-Dirac distribution where the first Dirac pulse is located at $x = 1$ with weight 0.25, while the second is at $x = 2$ with weight 0.75. The second case is the convolution of the same Gaussian curve with a uniform distribution between $x = 1$ and $x = 3$.

Note how the Q-scale tail fitting method, while still giving a good fit of the tails, generally underestimates the DJ and, to compensate for that, overestimates the variance of the Gaussian curves, and thus the RJ. This might give overly pessimistic results for TJ for very low probabilities (refer to Section 2.2.6 for a review of RJ, DJ, and TJ).

The reason for this behavior can be readily understood. The Q-scale method assumes that the best fit Gaussian curves are of the form of Equation 10.26, thus having an area equal to one. But this cannot hold in the presence of DJ, especially when the DJ component is significant. Indeed, in such cases, part of the area below the PDF must be attributed to DJ, so that the area accounted for by the Gaussian distributions (or, in other terms, by the RJ) must be less than 1. This is evident by looking at the top plot of Figures 10.5 and 10.6: the amplitude of the fitted Gaussian is too large to be realistic.

Recognizing these limitations, a new method was introduced, involving the estimation of the amplitude A of the Gaussian distribution. The basic theory can be traced back to the work of Popovici [125], addressing the problem of the extrapolation of bit error

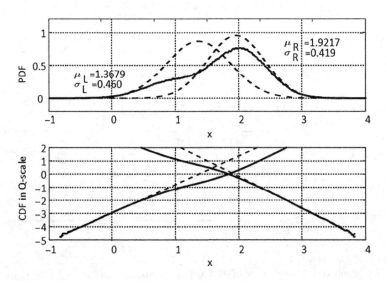

Figure 10.5 Tail fitting with Q-scale applied to a dual-Dirac DJ distribution. Top: PDF of the measured jitter (solid) and fitted Gaussian curves (dashed). Bottom: CDF of the measured jitter (solid) and fitted Gaussian curves (dashed) plotted in Q-scale.

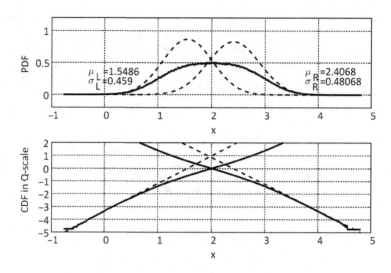

Figure 10.6 Tail fitting with Q-scale applied to a uniform DJ distribution. Top: PDF of the measured jitter (solid) and fitted Gaussian curves (dashed). Bottom: CDF of the measured jitter (solid) and fitted Gaussian curves (dashed) plotted in Q-scale.

rates in digital links in the presence of amplitude noise, and has been further developed recently (see, e.g., [126], [127]) for application in specialized test equipment.

The starting point is to recognize that the Gaussian curve to be fitted is equal to a Gaussian PDF multiplied by an amplitude factor A less than or equal to 1:

$$n_A(x) = A \cdot n(x) = \frac{A}{\sigma\sqrt{2\pi}} \exp\left[-\frac{(x-\mu)^2}{2\sigma^2}\right]. \tag{10.31}$$

As a result, the corresponding CDF is $N_A(x) = A \cdot N(x)$, where $N(x)$ is the CDF of a Gaussian distribution. It can be easily seen that application of the Q-scale transform to $N_A(x)/A$ will linearize the tail of the PDF under test and allow the extraction of the mean and variance:

$$-\sqrt{2}\,\mathrm{erfc}^{-1}\left[\frac{2N_A(x)}{A}\right] = \frac{x-\mu}{\sigma}. \tag{10.32}$$

The fact that CDF under test is divided by the amplitude A before applying the Q-scale transform gives this method the name of *normalized* Q-scale transform.

In the context of tail fitting, the amplitude A is one of the unknown parameters to be identified. The approach described by [125] is to divide the CDF under test $\Phi(x)$ by a variable factor k and then apply the Q-scale transform:

$$-\sqrt{2}\,\mathrm{erfc}^{-1}\left[\frac{2\Phi(x)}{k}\right]. \tag{10.33}$$

It can be readily understood that the Q-scale transformation will result in a straight line only if $k = A$. Figure 10.7 illustrates this concept with an example obtained with a Gaussian CDF of amplitude $A = 0.3$. The top graph shows PDF and CDF of the distribution, while the bottom graph shows the result of the application of the normalized

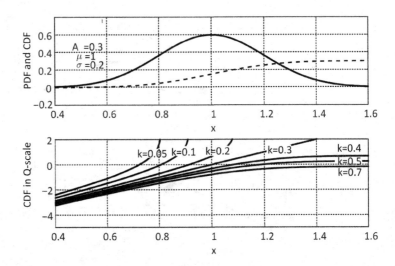

Figure 10.7 Linearization of the CDF of a Gaussian distribution, by use of the normalized Q-scale transformation. Top: PDF and CDF of a Gaussian distribution with $\mu = 1$, $\sigma = 0.2$, and $A = 0.3$. Bottom: CDF plot obtained by applying the normalized Q-scale transformation 10.33.

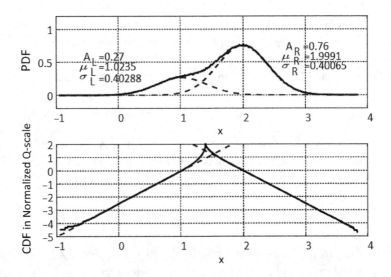

Figure 10.8 Tail fitting with normalized Q-scale applied to a dual-Dirac DJ distribution. Top: PDF of the measured jitter (solid) and fitted Gaussian curves (dashed). Bottom: CDF of the measured jitter (solid) and fitted Gaussian curves (dashed) plotted in normalized Q-scale.

Q-scale transform with different factors k. Only for $k = 0.3$ is the result a perfect straight line, and the mean and standard deviation of the Gaussian can be extracted from the slope and the intercept with the x-axis.

Using this result, tail fitting including estimation of amplitude A can be performed in the following steps. As before, first the CDF of the left and right tails are calculated based on the PDF or histogram of the measured data. Second, the left and right CDF

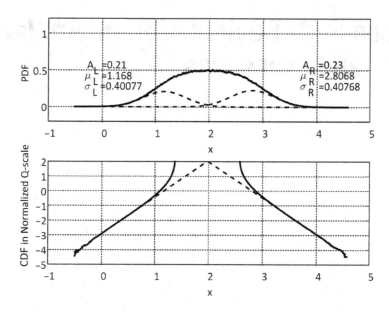

Figure 10.9 Tail fitting with normalized Q-scale applied to a uniform DJ distribution. Top: PDF of the measured jitter (solid) and fitted Gaussian curves (dashed). Bottom: CDF of the measured jitter (solid) and fitted Gaussian curves (dashed) plotted in normalized Q-scale.

are divided by a variable scaling factor k and the Q-scale transform of Equation 10.30 is applied. The result of the Q-scale transform is tested by varying the factor k, and the amplitude A is assumed to be equal to the k factor giving the best linear result. Third, a best linear fit is applied to the linearized CDFs and, finally, the values of the mean and sigma for the left and right tails are extracted from the fitted lines.

Figures 10.8 and 10.9 show the result of the procedure outlined above, applied to two cases of dual-Dirac and uniform DJ already presented before. It can be noted how the normalized Q-scale approach delivers a more realistic fitting of the tail of the distribution and more accurate results for the DJ as well as for the RJ components.

Section B.3 reports Matlab functions implementing the Q-scale and the normalized Q-scale methods.

Note that one of the biggest challenges is the selection of the range of the distribution over which to apply the tail fitting algorithm. Ideally we would like to take the most extreme parts of the PDF, so that the contamination of RJ due to DJ is at a minimum, but those parts are also the ones more subject to random variations and to inaccuracy due to the limited number of samples available. Following [128], the algorithms presented in Section B.3 evaluate the linearity of the transformed CDF over a variable x range, and finally the one giving the least error when fitted with a straight line is chosen. This is meant to be just one possible example of how to tackle this problem, and many others might be equivalent or superior.

Appendix A **Review of Random Variables and Processes**

This appendix reviews in a very compact form the basic concepts of random variables and random processes for the purpose of understanding jitter and phase noise. The goal is not to provide a thorough treatment, but rather to help the reader in recalling the definitions and the meanings of some of the most useful and common concepts. It is assumed that the reader is familiar with the basic concepts of probability. Although not strictly necessary, it is beneficial that the reader has been previously exposed to the theory of random variables and random processes, as can be found in a typical graduate university class. For further reading, or an in-depth analysis of the concepts, such specialized textbooks as [1], [13], [129], and [105] can be consulted.

A.1 Random Variables

A.1.1 Definition

Given an experiment with many possible outcomes, a *random variable X* is defined as a function that assigns a real number to each outcome of this experiment. In order to be called a random variable, the function must satisfy the following condition: the set of outcomes for which X is smaller than or equal to a given real number a, has a given, well-defined probability for any a.

As an example, let's take as an experiment the toss of a dice. The possible outcomes are each of the six faces of the dice. We can associate with each face the integer numbers 1 to 6, even if there are no numbers (or dots) on the faces of the dice (assuming that we can distinguish each of the faces of the dice somehow). This rule of association can be formalized as a function X which can assume the values $1, 2, 3, 4, 5, 6$. Given any real number a, the probability that $X < a$ is always well defined. For instance, the probability that $X \leq 3.4567$ is equal to the probability that the dice shows one of the faces associated with the numbers 1 to 3. The probability that $X \leq -1$ is obviously zero, since there is no possible outcome associated to numbers below 1. Since the condition stated in the above paragraph is satisfied, X can be called a random variable.

Note that the name "random variable" is, strictly speaking, a misnomer, since it refers to a function, not to a variable. However, in common practice the mechanism of associating a number with the outcomes of the experiment is often not defined explicitly, so that the property of being "random" is improperly associated with the value returned by the function X (hence the name "random variable"), rather than with the underlying experimental outcome. In many cases taken from engineering or physics, the result of the experiment is in itself a number, and the association of the result with a real number is not necessary. For instance, the measurement of the voltage noise of an electronic device at a given time is already in the form of a real number. This is why, very often, we identify with the term "random variable" the numerical result of the experiment itself.

A random variable is called *discrete* if it assumes values only within a set of distinct real numbers, while it is called *continuous* if it can assume any value in a continuous interval of the real axis.

A.1.2   Distribution and Density Functions

The *cumulative distribution function* (CDF) $F_X(a)$ of a random variable X is defined as the probability that $X \leq a$:

$$F_X(a) := P[X \leq a]. \tag{A.1}$$

It is easy to understand that the CDF of any random variable is a non-decreasing function of a and that $F_X(-\infty) = 0$, while $F_X(+\infty) = 1$. The CDF is often simply referred to as the *distribution function*.

The *probability density function* (PDF), or *density function* for short, $f_X(a)$ is defined as the derivative of the CDF:

$$f_X(a) := \frac{\partial F_X(a)}{\partial a}. \tag{A.2}$$

Recalling the definitions of derivative and of CDF, the PDF is equal to the probability of the random variable X assuming values in a small interval ϵ around a, divided by ϵ, for ϵ tending to 0.

The probability that the random variable X assumes values between a_1 and a_2 is equal to the difference of the CDF in a_1 and a_2, or, equivalently, to the integral of the PDF between a_1 and a_2:

$$P[a_1 < X \leq a_2] = F_X(a_2) - F_X(a_1) = \int_{a_1}^{a_2} f_X(v)\, dv. \tag{A.3}$$

From the above properties, it follows that any PDF is a nonnegative real function, defined over the whole real axis, and its integral from $-\infty$ to $+\infty$ is equal to 1.

In case of a continuous random variable, the CDF and PDF are continuous functions. For a discrete random variable the CDF is a stepwise increasing function, while the PDF is made of Dirac impulses, with area equal to the probability that the random variable assumes that specific value.

A.1.3   Expectation

The *expectation* of a random variable X is the probability-weighted average of X, and is denoted as $E[X]$. If X is a discrete random variable assuming only a finite number N of different values $a_1, \ldots, a_N$, then the expectation of X is:

$$E[X] := \sum_{i=1}^{N} a_i P[X = a_i]. \tag{A.4}$$

If X is a continuous random variable, the equation above becomes an integral:

$$E[X] := \int_{-\infty}^{+\infty} a f_X(a) \, da. \tag{A.5}$$

The most important property of the expectation is its linearity. Given two random variables X and Y and any two real numbers a and b, the expectation of the random variable obtained by combining linearly X and Y is equal to:

$$E[aX + bY] = aE[X] + bE[Y]. \tag{A.6}$$

In many practical cases a random variable X is transformed into another random variable Y by application of a function $g(\cdot)$, $Y = g(X)$. It can be shown that in these cases the expectation of the random variable Y can be computed as:

$$E[Y] = E[g(X)] = \int_{-\infty}^{+\infty} g(a) f_X(a) \, da. \tag{A.7}$$

This result is known as the *fundamental theorem of expectation*. In case of a discrete random variable the result can be written as:

$$E[g(X)] = \sum_{i=1}^{N} g(a_i) P[X = a_i]. \tag{A.8}$$

A notable case of function of a random variable is given by the *characteristic function* $\Phi_X(\omega)$, defined as:

$$\Phi_X(\omega) := E[e^{i\omega X}] = \int_{-\infty}^{+\infty} e^{i\omega a} f_X(a) \, da. \tag{A.9}$$

It can be seen that the characteristic function is the complex conjugate of the Fourier transform of the PDF.

A.1.4   Mean, Variance, and Higher-Order Moments

The *mean* of a random variable X is simply the expectation of that variable, $E[X]$, and is a measure of the center of the variable's PDF:

$$\mu_X := E[X]. \tag{A.10}$$

The *variance* is a measure of the spread of the PDF around the center point as is defined as:

$$\sigma_X^2 := E\left[(X - \mu_X)^2\right]. \tag{A.11}$$

It can be easily shown that $\sigma_X^2 = E[X^2] - \mu_X^2$. For a random variable with zero mean the variance is equal to the second moment $E[X^2]$. The squared root of the variance is called *standard deviation* and is denoted by σ_X.

In general, the quantity $E[X^n]$ is called the *n-th moment* of the random variable X and $E[(X - \mu_X)^n]$ is called the *n-th central moment*.

The characteristic function of a random variable can be used to compute the high-order moments by simple differentiation:

$$E[X^n] = \frac{1}{i^n} \left. \frac{\partial^n \Phi_X(\omega)}{\partial \omega^n} \right|_{\omega=0}. \tag{A.12}$$

A.1.5 Two Random Variables

Experiments in which two random variables X and Y are involved can be described by the *joint cumulative distribution function* $F_{XY}(a, b)$, defined as:

$$F_{XY}(a, b) := P[X \le a \text{ and } Y \le b]. \tag{A.13}$$

The *joint probability density function* $f_{XY}(a, b)$ is the partial derivative of the joint CDF:

$$f_{XY}(a, b) := \frac{\partial^2}{\partial a \, \partial b} F_{XY}(a, b). \tag{A.14}$$

From this definition it can be shown that:

$$F_{XY}(a, b) = \int_{-\infty}^{a} \int_{-\infty}^{b} f_{XY}(v, w) \, dv dw. \tag{A.15}$$

In general, the probability that the couple (X, Y) is in a region D of the bi-dimensional plane is given by the integral of the joint PDF over the region D:

$$P[(X, Y) \in D] = \int \int_D f_{XY}(v, w) \, dv dw. \tag{A.16}$$

From the joint distribution, the distributions of the single variable can be computed as:

$$\begin{aligned} F_X(a) &= F_{XY}(a, +\infty) \\ f_X(a) &= \int_{-\infty}^{+\infty} f_{XY}(a, b) \, db. \end{aligned} \tag{A.17}$$

A.1.6 Independent Random Variables

Two random variables X and Y are said to be *independent* if for any two events A and B:

$$P[X \in A \text{ and } Y \in B] = P[X \in A] \cdot P[Y \in B]. \tag{A.18}$$

In this case, the joint distribution functions are the product of the distribution functions of the single random variables:

$$\begin{aligned} F_{XY}(a, b) &= F_X(a)F_X(b) \\ f_{XY}(a, b) &= f_X(a)f_X(b). \end{aligned} \tag{A.19}$$

A.1.7   Expectation of a Function of Two Random Variables

Similar to the case of one random variable, two random variables X and Y can be transformed into another random variable Z by means of a function $g(\cdot,\cdot)$, $Z = g(X,Y)$. According to the original definition, the expectation of Z is given by:

$$\mathrm{E}\left[Z\right] = \int_{-\infty}^{+\infty} a f_Z(a)\,\mathrm{d}a \tag{A.20}$$

however it can be computed in a much easier way directly in terms of the function $g(\cdot,\cdot)$ and of the joint PDF $f_{XY}(a,b)$:

$$\mathrm{E}\left[g(X,Y)\right] = \int_{-\infty}^{+\infty} \int_{-\infty}^{+\infty} g(a,b) f_{XY}(a,b)\,\mathrm{d}a\,\mathrm{d}b. \tag{A.21}$$

A.1.8   Correlation and Covariance

Correlation and *covariance* are a measure of the interdependency of two random variables and are indicated as R_{XY} and C_{XY} respectively. They are defined respectively as the expectation of the product of X and Y, and of the product of $X - \mu_X$ and $Y - \mu_Y$

$$\begin{aligned} R_{XY} &:= \mathrm{E}\left[XY\right] \\ C_{XY} &:= \mathrm{E}\left[(X - \mu_X)(Y - \mu_Y)\right]. \end{aligned} \tag{A.22}$$

It is easy to show that:

$$C_{XY} = R_{XY} - \mu_X \mu_Y. \tag{A.23}$$

The *correlation coefficient* ρ_{XY} of two random variables is defined as:

$$\rho_{XY} := \frac{C_{XY}}{\sigma_X \sigma_Y} \tag{A.24}$$

and it can be shown to be always less than or equal to 1 in absolute value:

$$|\rho_{XY}| \leq 1. \tag{A.25}$$

Two random variables are called *uncorrelated* if their covariance is zero, or, based on the relationship above, if:

$$\mathrm{E}\left[XY\right] = \mathrm{E}\left[X\right]\mathrm{E}\left[Y\right]. \tag{A.26}$$

Two random variables are called *orthogonal* if their correlation is zero:

$$\mathrm{E}\left[XY\right] = 0. \tag{A.27}$$

It is easy to show that if two random variables are independent, then they are also uncorrelated.

A.2   Random Processes

A.2.1   Definition

While a random variable is a rule to associate a real number with each possible outcome of an experiment, strictly speaking a *random process* is a rule to associate with the experiment's outcome a *function* of time $x(t)$. Since the experiment can have multiple outcomes, in some cases also a non-countable number of outcomes, the random process defines a family of time functions associated with the experiment.

As in the case of the random variable, the results of many experiments in engineering and physics are already in the form of a function of time. For example, the position of a particle subject to Brownian motion, the output voltage noise of a resistor over time, the total number of particles emitted by a radioactive substance, and the number of persons in a line waiting to be served, can be expressed in terms of a real-time function. For this reason, in the common usage in engineering and physics, the term "random process" often identifies directly the result of the experiment over time.

Note that if t is fixed, $x(t)$ is a random variable equal to the state of the process at time t.

A.2.2   Classification of Random Processes

Random processes can be classified based on the nature of the variable t and on the nature of the values returned by the function $x(t)$. If t assumes values over the whole real axis, then $x(t)$ is called *continuous-time* process. If t assumes only integer values then the process is called *discrete-time* process. In the case of a discrete-time process, the process is typically indicated by x_n, where n can assume integer values. If the function $x(t)$ returns only a countable number of values, it is called *discrete-state*, otherwise it is called *continuous-state*.

As an example, the total number of particles emitted by a radioactive substance over time is a continuous-time discrete-state process. The process obtained my tossing a dice multiple times is a discrete-time discrete-state process. Notably, jitter on a clock is a discrete-time process, while the excess phase is a continuous-time one. Both of them are continuous-state.

A.2.3   Mean, Autocorrelation, and Autocovariance

The *mean* $\mu_x(t)$ of a random process $x(t)$ is defined as the expectation of the random variable $x(t)$, for each value of t:

$$\mu_x(t) := \mathrm{E}\left[x(t)\right] \tag{A.28}$$

and is a function of time t. The mean $\mu_x(t)$ can be considered as the average waveform of the process $x(t)$.

The *autocorrelation* $R_x(t_1, t_2)$ is the expectation of the product $x(t_1)x(t_2)$:

$$R_x(t_1, t_2) := \mathrm{E}\left[x(t_1)x(t_2)\right]. \tag{A.29}$$

The autocorrelation for $t_1 = t_2 = t$ is the average power of $x(t)$, or, in other terms, the second moment of the random variable $x(t)$:

$$R_x(t, t) := \mathrm{E}\left[x(t)^2\right].$$ (A.30)

If $x(t)$ has zero mean, then $R_x(t, t)$ is the variance of $x(t)$.

The *autocovariance* $C_x(t_1, t_2)$ is the expectation of the product of $x(t_1) - \mu_x(t_1)$ and $x(t_2) - \mu_x(t_2)$, or, in other terms, the covariance of the two random variables $x(t_1)$ and $x(t_2)$:

$$C_x(t_1, t_2) := \mathrm{E}\left[(x(t_1) - \mu_x(t_1))(x(t_2) - \mu_x(t_2))\right].$$ (A.31)

The autocovariance for $t_1 = t_2 = t$ is the variance of the random variable $x(t)$.

Autocorrelation and autocovariance are related to each other by:

$$C_x(t_1, t_2) = R_x(t_1, t_2) - \mu_x(t_1)\mu_x(t_2).$$ (A.32)

If a process has zero mean, autocorrelation and autocovariance are identical.

An alternative, and more useful, way of writing the autocorrelation and autocovariance is by using the lag variable $\tau = t_2 - t_1$:

$$R_x(t, \tau) := \mathrm{E}\left[x(t + \tau)x(t)\right].$$ (A.33)

A.2.4   Stationary Processes

A random process $x(t)$ is called *strict-sense stationary* if its statistical properties are invariant to a translation of the time axis. This means that the statistics of the process $x(t + u)$ are equal to the statistics of $x(t)$ for any u. In nature, many processes are strict sense stationary, if the environmental conditions of the experiment don't change over time. For example, the process consisting in an infinite sequence of dice tossing is a strict-sense stationary process. The noise produced by a resistor in an environment where temperature doesn't change is another example.

If only the mean and the autocorrelation (and thus also autocovariance) of the process are invariant to a translation of the time axis:

$$\begin{aligned}
\mu_x(t + u) &= \mu_x(t) \\
R_x(t_1, t_2) &= R_x(t_1 + u, t_2 + u)
\end{aligned}$$ (A.34)

then the process is called *wide-sense stationary*. Obviously a strict-sense stationary process is also stationary in the wide sense, but the converse is not generally true. From the equations above it is immediately apparent that for wide-sense stationary processes the mean is independent of time, $\mu_x(t) = \mu_x$, and the autocorrelation depends only on the difference between t_1 and t_2. For this class of processes the autocorrelation is typically expressed as a function of the single lag variable τ:

$$R_x(\tau) := \mathrm{E}\left[x(t + \tau)x(t)\right].$$ (A.35)

Note that for wide-sense stationary processes $\mathrm{E}\left[x(t)^2\right] = R_x(0)$; thus the power is equal to the value of the autocorrelation at zero lag, and is independent of time.

A.2.5 Wide-Sense Cyclostationary Processes

A random process $x(t)$ is called *wide-sense cyclostationary* if its mean and its auto-correlation are invariant to translation of the time axis by multiples of a given period T:

$$\mu_x(t + mT) = \mu(t)$$
$$R(t, \tau) = R(t + mT, \tau) \tag{A.36}$$

for any integer m.

It can be shown that if $x(t)$ is a wide-sense cyclostationary process with period T and u is a random variable independent from the process $x(t)$ and uniformly distributed between 0 and T, then the new process $y(t) = x(t - u)$ obtained by shifting the time origin by u is wide-sense stationary. This operation is called *phase randomization* of the cyclostationary process $x(t)$. The mean and autocorrelation of $y(t)$ turn out to be the time average of the mean and autocorrelation of the original process $x(t)$ over one period T:

$$m_y = \tfrac{1}{T} \int_0^T m_x(t)\, dt$$
$$R_y(\tau) = \tfrac{1}{T} \int_0^T R_x(t, \tau)\, dt. \tag{A.37}$$

Phase randomization is an acceptable method of reducing the cyclostationarity of a process to stationarity in the wide sense, in the case where the phase of the original process is either not known or not of interest.

A.2.6 Gaussian Processes

A random process $x(t)$ is called *Gaussian* or *Normal* if, for any integer n, and for any choice of n time samples $x(t_1) \ldots x(t_n)$, the random variable

$$Y = \sum_{i=1}^{n} a_i x(t_i) \tag{A.38}$$

is a Gaussian random variable for any value of the coefficients a_1 to a_n. An interesting property of Gaussian processes is that their complete statistical properties can be described by using only the mean and autocovariance, or autocorrelation. Gaussian processes are very common in nature and a very useful model to describe a variety of real-life applications.

A.2.7 Power Spectral Density and the Wiener–Khinchin Theorem

Assume that we have a wide-sense stationary random process $x(t)$. In order to find the energy distribution of this process over frequency, we calculate its Fourier transform over a limited period of time $[-T/2, +T/2]$:

$$X_T(f) := \int_{-T/2}^{+T/2} x(t) e^{-j2\pi ft}\, dt. \tag{A.39}$$

Using Parseval's theorem, the quantity:

$$\int_{-\infty}^{+\infty} |X_T(f)|^2 \, df \tag{A.40}$$

is the energy of the signal over the interval $[-T/2, +T/2]$, so that

$$\frac{1}{T}|X_T(f)|^2 \tag{A.41}$$

can be interpreted as distribution of power in the frequency domain. For each f this quantity is a random variable, since it is a function of the random process $x(t)$. The *power spectral density* (PSD) $S_x(f)$ is defined as the limit of the expectation of the expression above, for large T:

$$S_x(f) := \lim_{T \to \infty} \mathrm{E}\left[\frac{1}{T}|X_T(f)|^2\right]. \tag{A.42}$$

The *Wiener–Khinchin theorem* ensures that for well-behaved wide-sense stationary processes the limit exists and is equal to the Fourier transform of the autocorrelation:

$$\begin{aligned} S_x(f) &= \int_{-\infty}^{+\infty} R_x(\tau)e^{-j2\pi f\tau} \, d\tau \\ R_x(\tau) &= \int_{-\infty}^{+\infty} S_x(f)e^{j2\pi f\tau} \, df. \end{aligned} \tag{A.43}$$

An easy consequence of the Wiener–Khinchin theorem is that the total power of the signal is equal to the autocorrelation in zero, and is given by the integral of the PSD over the whole frequency axis:

$$R_x(0) = \int_{-\infty}^{+\infty} S_x(f) \, df. \tag{A.44}$$

A.2.8   Engineering Definitions of the Power Spectral Density

In the practice of engineering, it has become customary to use slightly different variants of the PSD definition, depending on the particular application or research field. The most common ones are: one-sided PSD, two-sided PSD, single-sideband (SSB) PSD and double-sideband (DSB) PSD. Since these are often sources of confusion and misunderstanding, this subsection aims to clarify their meaning in accordance with their most widespread usage (see, e.g., [24], [130], [26] p. 5, [25]).

- **Two-Sided PSD, $S_x(f)$**: this is a synonym of the PSD defined as the Fourier Transform of the autocorrelation.
- **One-Sided PSD, $S_x'(f)$**: this is a variant derived from the two-sided PSD by considering only the positive frequency semi-axis. To conserve the total power, the value of the one-sided PSD is twice that of the two-sided PSD:

$$S_x'(f) := \begin{cases} 0 & \text{if } f < 0 \\ S_x(f) & \text{if } f = 0 \\ 2S_x(f) & \text{if } f > 0. \end{cases} \tag{A.45}$$

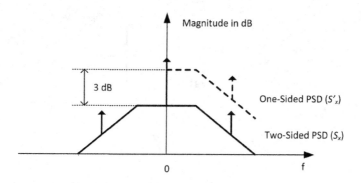

Figure A.1 Illustration of the definition of One-Sided and Two-Sided PSD.

Note that the one-sided PSD definition makes sense only if the two-sided is an even function of f. This is always the case if $x(t)$ is a real process, but might not be true if $x(t)$ is complex. The total power of the process can be calculated by integrating $S'_x(f)$ from zero to infinity. The relationship between one-sided and two-sided PSD is illustrated in Figure A.1.

If $S'_x(f)$ is even symmetrical around a positive frequency f_0, then two additional definitions can be adopted:

- **Single-Sideband PSD**, $S_{SSB,x}(f)$: this is obtained from $S'_x(f)$ by moving the origin of the frequency axis to f_0:

$$S_{SSB,x}(f) := S'_x(f_0 + f). \qquad (A.46)$$

By doing so, the frequency f in the SSB PSD assumes the meaning of the offset frequency from f_0. This concept is particularly useful for describing phase or amplitude modulation schemes in wireless communications, where f_0 is the carrier frequency. Note that there is no difference in the values of the one-sided versus the SSB PSD; it is just a pure translation on the frequency axis. The total power of the process can be calculated by integrating the SSB PSD from minus infinity to plus infinity.

- **Double-Sideband PSD**, $S_{DSB,x}(f)$: this is a variant of the SSB PSD obtained by considering only the positive frequency semi-axis. As in the case of the one-sided PSD, to conserve total power, the value of the DSB PSD is twice that of the SSB:

$$S_{DSB,x}(f) := \begin{cases} 0 & \text{if } f < 0 \\ S_{SSB,x}(f) & \text{if } f = 0 \\ 2S_{SSB,x}(f) & \text{if } f > 0. \end{cases} \qquad (A.47)$$

Note that the DSB definition makes sense only if the SSB PSD is even symmetrical around zero. In the case of a modulated carrier, this holds if the modulation affects only the phase. If additionally the amplitude is modulated, the SSB could by asymmetrical, and the DSB definition cannot be correctly applied. The relation among DSB, SSB, and one-sided PSD is illustrated in Figure A.2.

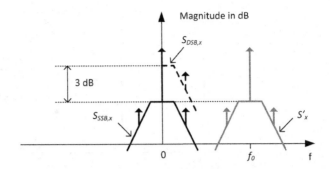

Figure A.2 Illustration of the definition of Single-Sideband and Double-Sideband PSD.

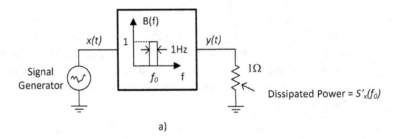

a)

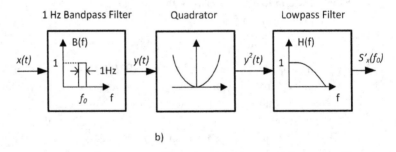

b)

Figure A.3 Physical meaning of the PSD: (a) experiment, (b) mathematical model.

A.2.9   The Physical Meaning of the Power Spectral Density

In this section we want to show that the one-sided PSD $S'_x(f)$ represents the average power of the signal x measured over a 1Hz band around the frequency f. Let us assume the signal x represents a physical voltage delivered by a voltage source over a 1Ohm resistor. The instantaneous power delivered by the source and dissipated in the resistor is equal to $P(t) = x^2(t)$.

We are interested to know how much of the total average power is delivered by frequency components in the range $f_0 - 1/2$ Hz to $f_0 + 1/2$ Hz. In a thought experiment, we can apply an ideal band-pass filter $B(f)$ between the signal source and the resistor and then measure the dissipated power on the resistor (see [1]). This experiment is illustrated in Figure A.3 together with its mathematical description. The measurement process is modeled as a low-pass filter $H(f)$ with unity gain at zero frequency. The instantaneous power delivered to the resistor, indicated as $P(f_0, 1Hz)$, is equal to $y^2(t)$. Using basic concepts of linear systems (see Section A.2.10), its average value is equal to the autocorrelation of the process y at zero:

$$\mathrm{E}\left[P(f_0, 1Hz)\right] = H(0)\mathrm{E}\left[y^2(t)\right] = \mathrm{E}\left[y^2(t)\right] = R_y(0) \tag{A.48}$$

since the low-pass filter used has gain 1 at zero frequency. From the definition of two-sided PSD we have:

$$R_y(0) = \int_{-\infty}^{+\infty} S_y(f)df = \int_{-\infty}^{+\infty} S_x(f)B^2(f)df. \tag{A.49}$$

Given that the band-pass filter B has value 1 on the intervals $[f_0 - 1/2, f_0 + 1/2]$ and $[-f_0 - 1/2, -f_0 + 1/2]$ and zero otherwise, and considering $S_x(f)$ almost constant on those intervals:

$$R_y(0) = \int_{-f_0-1/2}^{-f_0+1/2} S_x(f)df + \int_{f_0-1/2}^{f_0+1/2} S_x(f)df = 2S_x(f_0) = S_x'(f_0). \tag{A.50}$$

So the one-sided PSD at frequency f_0 is equal to the power physically dissipated by the resistor in the described experiment due to the frequency components of the voltage signal in a 1Hz bandwidth around f_0:

$$S_x'(f_0) = \mathrm{E}\left[P(f_0, 1Hz)\right]. \tag{A.51}$$

A.2.10   Random Processes and Linear Systems

Consider a random process $y(t)$ obtained as the response of a time-invariant linear system L, with impulse response $h(t)$ and frequency response $H(f)$, to an input random process $x(t)$:

$$y(t) = L\left[x(t)\right]. \tag{A.52}$$

Since the expectation is a linear operator, the mean of $y(t)$ is equal to the response of the system to the mean of $x(t)$, independent of the matter of whether $x(t)$ is stationary or not:

$$\mathrm{E}\left[y(t)\right] = L\left[\mathrm{E}\left[x(t)\right]\right] \tag{A.53}$$

$$\mu_y(t) = \int_{-\infty}^{+\infty} \mu_x(t - \tau)h(\tau)\,d\tau. \tag{A.54}$$

The autocorrelation of $y(t)$, $R_y(t_1, t_2)$ is obtained from the autocorrelation of $x(t)$ by applying the linear transformation on both time arguments:

$$R_y(t_1, t_2) = \int_{-\infty}^{+\infty}\int_{-\infty}^{+\infty} R_x(t_1 - \tau_1, t_2 - \tau_2)h(\tau_1)h(\tau_2)\,d\tau_1 d\tau_2. \tag{A.55}$$

In particular, if $x(t)$ is wide-sense stationary then:

$$\mu_y = \mu_x \int_{-\infty}^{+\infty} h(t)\,dt = \mu_x H(0) \tag{A.56}$$

and

$$R_y(\tau) = \int_{-\infty}^{+\infty} R_x(\tau - \tau_1) c_h(\tau_1) \, d\tau_1 \tag{A.57}$$

where:

$$c_h(\tau_1) := \int_{-\infty}^{+\infty} h(\tau_1 + \tau_2) h^*(\tau_2) \, d\tau_2 \tag{A.58}$$

and, in terms of PSD:

$$S_y(f) = S_x(f) |H(f)|^2. \tag{A.59}$$

A.2.11   Ergodicity

The most important statistical properties of a random process $x(t)$ are obtained by applying the expectation operator to some functions of the process itself. For instance, the mean is obtained as $\mathrm{E}[x(t)]$, and the autocorrelation as $\mathrm{E}[x(t_1)x(t_2)]$. The expectation returns the probability-weighted average of the specific function at that specific time *over all possible realizations* of the process. As an example, the mean $\mu(t) = \mathrm{E}[x(t)]$ represents the average value of the process $x(t)$ at the specific time t, over all possible realizations of the process itself. In many real practical cases, though, data from many realizations of the process are not available. On the contrary, often only data from one of them are known. The problem of estimating the statistical parameter of a process from a single realization is therefore central in the theory of random processes.

A straightforward way of proceeding is to replace the expectation operator with the time average. Considering again the mean as an example, we could argue that the time average:

$$\frac{1}{2T} \int_{-T}^{+T} x(t) \, dt \tag{A.60}$$

is a good estimation for the expectation of the process, but this is not always the case.

A process is called *ergodic*, if its statistical properties can be reliably estimated by replacing the expectation operator with a time-average over any one of its realizations. In particular, it is called *mean-ergodic* or *correlation-ergodic*, if mean or correlation can be obtained by time-averages.

It is obvious that if the statistical properties change with time, a process cannot be ergodic. Therefore, a necessary condition for mean- and correlation-ergodicity is that the process is wide-sense stationary. However, this condition is not sufficient: ergodic processes are a subset of wide-sense stationary processes.

Stating the necessary and sufficient conditions for ergodicity is beyond the scope of this short review. However, one sufficient condition is particularly simple and powerful. It can be shown that if the autocovariance of a process $C_x(\tau)$ tends to zero for large τ (process asymptotically uncorrelated), then the process is mean-ergodic. The same conditions do not imply that the process is also autocorrelation-ergodic, though, except if the process is a Gaussian process. An asymptotically uncorrelated Gaussian process is both mean- and autocorrelation-ergodic.

Appendix B Matlab Code for Jitter Generation and Analysis

B.1 Generation of Jitter

B.1.1 Flat Phase Noise Profile

The following Matlab code offers a possible implementation of the described procedure in the case of a 1GHz clock with a flat phase noise level at -110dBc/Hz:

```
F0=1e9;
L0=-110;
npoints=1e6;
sigma=sqrt(10^(L0/10)/F0)/(2*pi);
t_id=1/F0*(0:npoints-1);
j=sigma*randn(1,npoints);
t=t_id+j;
```

where the function `randn` generates normally distributed random numbers with variance equal to 1.

B.1.2 Low-Pass, High-Pass, or Band-Pass Phase Noise Profiles

The following code describes the generation of jitter samples with a simple PLL spectrum (inband phase noise = -110dBc/Hz and 3dB frequency = 1MHz, carrier frequency = 1 GHz), using the Matlab built-in functions `butter` and `filter`.

```
F0=1e9;
L0=-110;
f3dB=1e6;
npoints=1e7;
sigma=sqrt(10^(L0/10)/F0)/(2*pi)
t_id=1/F0*(0:npoints-1);
j=sigma*randn(1,npoints);
[B,A] = butter(1,2*f3dB/F0);
j_filtered=filter(B,A,j);
t=t_id+j_filtered;
```

B.1.3 $1/f^2$ Phase Noise Profile

The following code describes the generation of jitter samples with a $1/f^2$ profile and -110dBc/ Hz ant 5MHz offset frequency for a 1GHz clock.

```
-----------------------------------------------------------------
F0=1e9;
L1=-110;
f1=5e6;
npoints=1e6;
sigma_l=(f1/F0)*sqrt(10^(L1/10)/F0)
t_id=1/F0*(0:npoints-1);
l=sigma_l*randn(1,npoints);
j=cumsum(l);
t=t_id+j;
-----------------------------------------------------------------
```

B.1.4 $1/f$ Phase Noise Profiles

The following code generates a $1/f$ phase noise profile with -60dBc/ Hz at 1kHz on a 1GHz carrier in the frequency range from 10Hz to 100MHz.

```
-----------------------------------------------------------------
F0=1e9;
fmin=1e1;
fmax=1e8;
L1=-60;
f1=1e3;
npoints=1e8;
t_id=1/F0*(0:npoints-1);

gamma=abs(sqrt(10)/(1+j*10)+1/(1+j)+sqrt(10)/(10+j))^2;
sigma_l=sqrt(10^(L1/10)*f1/(F0*fmin*gamma*4*pi^2));
l=sigma_l*randn(1,npoints);

fcorner=logspace(log10(fmin),log10(fmax),8);
[B1,A1] = butter(1,2*fcorner(1)/F0);
[B2,A2] = butter(1,2*fcorner(2)/F0);
[B3,A3] = butter(1,2*fcorner(3)/F0);
[B4,A4] = butter(1,2*fcorner(4)/F0);
[B5,A5] = butter(1,2*fcorner(5)/F0);
[B6,A6] = butter(1,2*fcorner(6)/F0);
[B7,A7] = butter(1,2*fcorner(7)/F0);
[B8,A8] = butter(1,2*fcorner(8)/F0);
```

```
l1_filtered=filter(B1,A1,l);
l2_filtered=filter(B2,A2,l/sqrt(10));
l3_filtered=filter(B3,A3,l/sqrt(10)^2);
l4_filtered=filter(B4,A4,l/sqrt(10)^3);
l5_filtered=filter(B5,A5,l/sqrt(10)^4);
l6_filtered=filter(B6,A6,l/sqrt(10)^5);
l7_filtered=filter(B7,A7,l/sqrt(10)^6);
l8_filtered=filter(B8,A8,l/sqrt(10)^7);

j=l1_filtered+l2_filtered+l3_filtered+l4_filtered+...
l5_filtered+l6_filtered+l7_filtered+l8_filtered;
t=t_id+j;
```

B.1.5 $1/f^3$ Phase Noise Profiles

The following code generates a $1/f^3$ noise profile with -80dBc/Hz at 10kHz on a 1GHz carrier.

```
F0=1e9;
fmin=1e1;
fmax=1e8;
L1=-80;
f1=10e3;
npoints=1e7;
t_id=1/F0*(0:npoints-1);

gamma=abs(sqrt(10)/(1+j*10)+1/(1+j)+sqrt(10)/(10+j))^2;
sigma_l=sqrt(10^(L1/10)*f1^3/(F0^3*fmin*gamma));
l=sigma_l*randn(1,npoints);

fcorner=logspace(log10(fmin),log10(fmax),8);
[B1,A1] = butter(1,2*fcorner(1)/F0);
[B2,A2] = butter(1,2*fcorner(2)/F0);
[B3,A3] = butter(1,2*fcorner(3)/F0);
[B4,A4] = butter(1,2*fcorner(4)/F0);
[B5,A5] = butter(1,2*fcorner(5)/F0);
[B6,A6] = butter(1,2*fcorner(6)/F0);
[B7,A7] = butter(1,2*fcorner(7)/F0);
[B8,A8] = butter(1,2*fcorner(8)/F0);

l1_filtered=filter(B1,A1,l);
l2_filtered=filter(B2,A2,l/sqrt(10));
l3_filtered=filter(B3,A3,l/sqrt(10)^2);
```

```
l4_filtered=filter(B4,A4,1/sqrt(10)^3);
l5_filtered=filter(B5,A5,1/sqrt(10)^4);
l6_filtered=filter(B6,A6,1/sqrt(10)^5);
l7_filtered=filter(B7,A7,1/sqrt(10)^6);
l8_filtered=filter(B8,A8,1/sqrt(10)^7);

j1=l1_filtered+l2_filtered+l3_filtered+l4_filtered+...
l5_filtered+l6_filtered+l7_filtered+l8_filtered;
j=cumsum(j1);
t=t_id+j;
```

B.1.6   More Complex Phase Noise Profiles

The following code describes the generation of a phase noise profile featuring a $1/f^3$ region, a $1/f^2$ region, a PLL-like spectrum, and finally a flat noise floor.

```
F0=1e9;
fmin=1e1;
fmax=1e8;

L1_flicker=-80;
f1_flicker=3e3;

L1_white=-100;
f1_white=30e3;

L0_pll=-100;
f3dB_pll=2e6;

L0_flat=-130;

npoints=1e8;
t_id=1/F0*(0:npoints-1);

gamma=abs(sqrt(10)/(1+j*10)+1/(1+j)+sqrt(10)/(10+j))^2;
sigma_l_flicker=sqrt(10^(L1_flicker/10)*f1_flicker^3...
/(F0^3*fmin*gamma));
l_flicker=sigma_l_flicker*randn(1,npoints);

fcorner=logspace(log10(fmin),log10(fmax),8);
[B1,A1] = butter(1,2*fcorner(1)/F0);
[B2,A2] = butter(1,2*fcorner(2)/F0);
[B3,A3] = butter(1,2*fcorner(3)/F0);
[B4,A4] = butter(1,2*fcorner(4)/F0);
```

```
[B5,A5] = butter(1,2*fcorner(5)/F0);
[B6,A6] = butter(1,2*fcorner(6)/F0);
[B7,A7] = butter(1,2*fcorner(7)/F0);
[B8,A8] = butter(1,2*fcorner(8)/F0);

l1_filtered=filter(B1,A1,l_flicker);
l2_filtered=filter(B2,A2,l_flicker/sqrt(10));
l3_filtered=filter(B3,A3,l_flicker/sqrt(10)^2);
l4_filtered=filter(B4,A4,l_flicker/sqrt(10)^3);
l5_filtered=filter(B5,A5,l_flicker/sqrt(10)^4);
l6_filtered=filter(B6,A6,l_flicker/sqrt(10)^5);
l7_filtered=filter(B7,A7,l_flicker/sqrt(10)^6);
l8_filtered=filter(B8,A8,l_flicker/sqrt(10)^7);

j1_flicker=l1_filtered+l2_filtered+l3_filtered+l4_filtered+...
l5_filtered+l6_filtered+l7_filtered+l8_filtered;
j_flicker=cumsum(j1_flicker);

sigma_l_white=(f1_white/F0)*sqrt(10^(L1_white/10)/F0);
l_white=sigma_l_white*randn(1,npoints);
j_white=cumsum(l_white);

sigma_l_pll=sqrt(10^(L0_pll/10)/F0)/(2*pi);
l_pll=sigma_l_pll*randn(1,npoints);
[B,A] = butter(1,2*f3dB_pll/F0);
j_pll=filter(B,A,l_pll);

sigma_j_flat=1/(2*pi)*sqrt(10^(L0_flat/10)/F0);
j_flat=sigma_j_flat*randn(1,npoints);

t=t_id+j_flicker+j_white+j_pll+j_flat;
```

B.2   Analysis of Jitter

The following code describes a Matlab function, which delivers the main jitter parameters and the phase noise profile starting from the array of transition instants t.

```
function [F_id,tj_per_pp,tj_per_rms,tj_c2c_pp,tj_c2c_rms,...
          tj_rms,tj_pp,fPHN,PHN] = f_extract_jitter_phn(t)

%% Compute Period Jitter
T=diff(t);
% Peak to Peak Period Jitter
```

```matlab
tj_per_pp=max(T)-min(T);
% Rms Period Jitter
tj_per_rms=std(T);
% Peak to Peak Cycle to Cycle Jitter
tj_c2c_pp=max(diff(T))-min(diff(T));
% Rms Cycle to Cycle Jitter
tj_c2c_rms=std(diff(T));

%% Compute least squares line
n=length(t);
i=(0:n-1);
T_id=(sum(i.*t)-(n-1)/2*sum(t))/((n^2-1)*n/12);
t_off=sum(t)/n-T_id*(n-1)/2;
F_id=1/T_id;

%% Absolute jitter
t_id=t_off+(0:T_id:(n-1)*T_id);
tj=t-t_id;
tj=tj-mean(tj);
% Standard Deviation Absolute Jitter
tj_rms=std(tj);
% Peak to Peak Absolute Jitter
tj_pp=max(tj)-min(tj);

%% Compute Phase Noise
phi_e=tj*2*pi/T_id;
npsd=2^(nextpow2(length(phi_e)/4)-1);  .
Fpsd=1/(npsd*T_id);
fPHN=0:Fpsd:Fpsd*(floor(npsd/2)-1);
nwind=32;
w=hann(npsd,'periodic')';
U2=sum(w.^2);
I=zeros(1,npsd);
for i=1:nwind
start=1+floor((n-npsd)/(nwind-1))*(i-1);
stop=start+npsd -1;
    xtmp=phi_e(start:stop);
v=xtmp.*w;
    I=I+abs(fft(v,npsd)).^2;
end
PHN=I/(nwind*U2/T_id);
PHN=PHN(1:floor(npsd/2));
```

B.3   Tail Fitting

The following code describes a Matlab function, which performs tail fitting according to the Q-scale methodology, starting from the PDF of the jitter, defined over the *x*-axis. It returns the mean and sigma values of the Gaussian curves fitting the left and right tails of the PDF.

```
-------------------------------------------------------------
function [mu_L,sigma_L,mu_R,sigma_R] = ...
    f_tail_fitting_q_scale(x, pdf, fitbins_strt, fitbins)
% x = x-axis over which the PDF is defined
% pdf = PDF of the jitter
% fitbins = array defining the length of bins used for tail fitting
% fitbins_strt = starting bin for tail fitting

%% build CDF
deltax=x(2)-x(1);
cdf_left=[0,cumsum(pdf)]*deltax;
cdf_right=[0,fliplr(cumsum(fliplr(pdf)))]*deltax;

%% fit left CDF
err_best=1;
for i_nbins=1:length(fitbins)
    i_nbins
    fit_interval=[fitbins_strt:fitbins_strt+fitbins(i_nbins)];
    cdf_left_q_scale=-sqrt(2)*erfcinv(2*cdf_left);
    c(i_nbins,:)=polyfit(x(fit_interval),...
                    cdf_left_q_scale(fit_interval),1);
    fitted=x(fit_interval)*c(i_nbins,1)+c(i_nbins,2);
    err(i_nbins)=std(cdf_left_q_scale(fit_interval)-fitted);
    if err(i_nbins)< err_best;
        err_best=err(i_nbins);
        i_nbins_best=i_nbins;
    end
end
sigma_L=1/c(i_nbins_best,1)
mu_L=-c(i_nbins_best,2)*sigma_L

%% fit right CDF
err_best=1;
for i_nbins=1:length(fitbins)
    fit_interval=[length(cdf_right)-fitbins_strt-fitbins(i_nbins)...
        :length(cdf_right)-fitbins_strt];
    cdf_right_q_scale=-sqrt(2)*erfcinv(2*cdf_right);
```

```
        c(i_nbins,:)=polyfit(x(fit_interval),...
                        cdf_right_q_scale(fit_interval),1);
        fitted=x(fit_interval)*c(i_nbins,1)+c(i_nbins,2);
        err(i_nbins)=std(cdf_right_q_scale(fit_interval)-fitted);
        if err(i_nbins)< err_best;
            err_best=err(i_nbins);
            i_nbins_best=i_nbins;
        end
end
sigma_R=-1/c(i_nbins_best,1)
mu_R=c(i_nbins_best,2)*sigma_R
```

The following code describes a Matlab function which performs tail fitting according to the normalized Q-scale methodology, starting from the PDF of the jitter, defined over the *x*-axis. It returns the amplitude, mean, and sigma values of the Gaussian curves fitting the left and right tails of the PDF.

```
function [A_L,mu_L,sigma_L,A_R, mu_R,sigma_R] = ...
    f_tail_fitting_norm_q_scale(x, pdf, fitbins_strt, fitbins)
% x = x-axis over which the PDF is defined
% pdf = PDF of the jitter
% fitbins = array defining the length of bins used for tail fitting
% fitbins_strt = starting bin for tail fitting


%% build CDF
deltax=x(2)-x(1);
cdf_left=[0,cumsum(pdf)]*deltax;
cdf_right=[fliplr(cumsum(fliplr(pdf))),0]*deltax;


k=[0.01:0.01:1];


%% fit left CDF
err_best=1;
for i_nbins=1:length(fitbins)
    fit_interval=[fitbins_strt:fitbins_strt+fitbins(i_nbins)];
    for i_k=1: length(k)
        cdf_left_q_scale=-sqrt(2)*erfcinv(2/k(i_k)*cdf_left);
        c(i_k,i_nbins,:)=polyfit(x(fit_interval),...
                        cdf_left_q_scale(fit_interval),1);
        fitted=x(fit_interval)*c(i_k,i_nbins,1)+c(i_k,i_nbins,2);
        err(i_k,i_nbins)=std(cdf_left_q_scale(fit_interval)-fitted);
        if err(i_k,i_nbins)< err_best;
            err_best=err(i_k,i_nbins);
```

```
                i_k_best=i_k;
                i_nbins_best=i_nbins;
            end
        end
end
sigma_L=1/c(i_k_best,i_nbins_best,1);
mu_L=-c(i_k_best,i_nbins_best,2)*sigma_L;
A_L=k(i_k_best);

%% fit right CDF
err_best=1;
for i_nbins=1:length(fitbins)
    fit_interval=[length(cdf_right)-fitbins_strt-fitbins(i_nbins)...
    :length(cdf_right)-fitbins_strt];
    for i_k=1: length(k)
        cdf_right_q_scale=-sqrt(2)*erfcinv(2/k(i_k)*cdf_right);
        c(i_k,i_nbins,:)=polyfit(x(fit_interval),...
                                 cdf_right_q_scale(fit_interval),1);
        fitted=x(fit_interval)*c(i_k,i_nbins,1)+c(i_k,i_nbins,2);
        err(i_k,i_nbins)=std(cdf_right_q_scale(fit_interval)-fitted);
        if err(i_k,i_nbins)< err_best;
            err_best=err(i_k,i_nbins);
            i_k_best=i_k;
            i_nbins_best=i_nbins;
        end
    end
end
sigma_R=-1/c(i_k_best,i_nbins_best,1);
mu_R=c(i_k_best,i_nbins_best,2)*sigma_R;
A_R=k(i_k_best);
------------------------------------------------------------------
```

Bibliography

[1] A. Papoulis and S. U. Pillai, *Probability, Random Variables and Stochastic Processes*, 4th ed. McGraw-Hill, 2002.

[2] "ITU-T Recommendation G.810: Definitions and Terminology for Synchronization Networks," *International Telecommunication Union (ITU)*, 1996.

[3] S. Bregni, "Measurement of Maximum Time Interval Error for Telecommunications Clock Stability Characterization," *IEEE Transactions on Instrumentation and Measurement*, vol. 45, no. 5, pp. 900–906, Oct. 1996.

[4] "Definition of Skew Specifications for Standard Logic Devices," *JEDEC Standard JESD65B*, 2003.

[5] T. C. Weigandt, B. Kim, and P. R. Gray, "Analysis of Timing Jitter in CMOS Ring Oscillators," *Proceedings of the International Symposium on Circuits and Systems (ISCAS)*, vol. 4, pp. 27–30, 1994.

[6] A. Demir, A. Mehrotra, and J. Roychowdhury, "Phase Noise in Oscillators: A Unifying Theory and Numerical Methods for Characterization," *IEEE Transactions on Circuits and Systems I: Regular Papers*, vol. 47, pp. 655–674, May 2000.

[7] "Understanding SYSCLK Jitter," Freescale, Tech. Rep., Feb. 2010, application Note AN4056.

[8] E. Pineda, "Clocks Basics in 10 Minutes or Less," Texas Instruments, Tech. Rep., 2010, webinar.

[9] D. C. Lee, "Analysis of Jitter in Phase-Locked Loops," *IEEE Transactions on Circuits and Systems II: Express Briefs*, vol. 49, no. 11, pp. 704–711, Nov. 2002.

[10] F. Herzel and B. Razavi, "A Study of Oscillator Jitter Due to Supply and Substrate Noise," *IEEE Transactions on Circuits and Systems II: Express Briefs*, vol. 46, pp. 56–62, Jan. 1999.

[11] W. Kester, *The Data Conversion Handbook*. Newnes, 2004, Analog Devices Series.

[12] ——, "Aperture Time, Aperture Jitter, Aperture Delay Time – Removing the Confusion," Analog Devices, Tech. Rep., 2009, tutorial MT-007.

[13] R. E. Walpole, R. H. Myers, S. L. Myers, and K. Ye, *Probability and Statistics for Engineers and Scientists*. Prentice-Hall, 2007.

[14] B. Ham, "Fibre Channel – Methodologies for Jitter and Signal Quality Specification," *International Committee for Information Technology Standardization (INCITS)*, Jun. 2005.

[15] M. P. Li, *Jitter, Noise, and Signal Integrity at High-Speed*. Prentice-Hall, 2008.

[16] G. Hänsel, K. Stieglbauer, G. Schulze, and J. Moreira, "Implementation of an Economic Jitter Compliance Test for a Multi-Gigabit Device on ATE," *IEEE International Test Conference (ITC04)*, pp. 1303–1312, Oct. 2004.

[17] D. Hong and K.-T. Cheng, "An Accurate Jitter Estimation Technique for Efficient High Speed I/O Testing," *IEEE Asian Test Symposium (ATS07)*, vol. 224–229, Oct. 2007.

[18] J. L. Huang, "A Random Jitter Extraction Technique in the Presence of Sinusoidal Jitter," *IEEE Asian Test Symposium (ATS06)*, vol. 318–326, Nov. 2006.

[19] M. Li, J. Wilstrup, R. Jessen, and D. Petrich, "A New Method for Jitter Decomposition Through Its Distribution Tail Fitting," *IEEE International Test Conference (ITC99)*, pp. 788–794, Sep. 1999.

[20] R. Stephens, "Separation of Random and Deterministic Components of Jitter," *US Patent 7 149 638*, Dec. 2006.

[21] ——, "Separation of a Random Component of Jitter and a Deterministic Component of Jitter," *US Patent 7 191 080*, Mar. 2007.

[22] S.Wisetphanichkij and K. Dejhan, "Jitter Decomposition by Derivatived Gaussian Wavelet Transform," *IEEE International Symposium on Communication and Information Technology (ISCIT04)*, vol. 2, pp. 1160–1165, Oct. 2004.

[23] S. Erb and W. Pribyl, "Design and Performance Considerations for an On-Chip Jitter Analysis System," *IEEE International Symposium on Circuits and Systems (ISCAS)*, pp. 3969–3972, May 2010.

[24] "IEEE Standard Definitions of Physical Quantities for Fundamental Frequency and Time Metrology – Random Instabilities," *IEEE Std 1139–2008*, Feb. 2008.

[25] A. Hajimiri and T. H. Lee, *The Design of Low Noise Oscillators*. Kluwer Academic Publishers, 1999.

[26] A. Lacaita, S. Levantino and C. Samori, *Integrated Frequency Synthesizers for Wireless Systems*. Cambridge University Press, 2007.

[27] G. Marzin, S. Levantino, C. Samori and A. Lacaita, "A Background Calibration Technique to Control Bandwidth in Digital PLLs," *Proceedings of the International Solid State Circuit Conference*, pp. 54–55, Feb. 2014.

[28] E. Temporiti, C. Weltin-Wu, D. Baldi, R. Tonietto, and F. Svelto, "A 3GHz Fractional All-Digital PLL With a 1.8MHz Bandwidth Implementing Spur Reduction Techniques," *IEEE Journal of Solid State Circuits*, vol. 44, no. 3, pp. 824–834, Mar. 2009.

[29] C. Samori, A. L. Lacaita, A. Zanchi, and F. Pizzolato, "Experimental Verification of the Link Between Timing Jitter and Phase Noise," *Electronics Letters*, vol. 34, no. 21, pp. 2024–2025, Oct. 1998.

[30] M. Abramowitz and I. Stegun, Eds., *Handbook of Mathematical Functions with Formulas, Graphs, and Mathematical Tables*. Dover, 1965.

[31] C. Liu and J. A. McNeill, "Jitter in Oscillators with 1/f Noise Sources," *IEEE International Symposium on Circuits and Systems (ISCAS)*, vol. 1, pp. 773–776, 2004.

[32] A. Demir, "Computing Timing Jitter from Phase Noise Spectra for Oscillators and Phase-Locked Loops with White and $1/f$ Noise," *IEEE Transactions on Circuits and Systems I: Regular Papers*, vol. 53, no. 9, pp. 1869–1884, Sep. 2000.

[33] A. Abidi, "Phase Noise and Jitter in CMOS Ring Oscillators," *IEEE Journal of Solid State Circuits*, pp. 1803–1816, Aug. 2006.

[34] W. Sansen, *Analog Design Essentials*. Springer, 2006.

[35] T. Pialis and K. Phang, "Analysis of Timing Jitter in Ring Oscillators Due to Power Supply Noise," *IEEE International Symposium on Circuits and Systems (ISCAS)*, vol. 1, pp. 685–688, 2003.

[36] A. Strak and H. Tenhunen, "Analysis of Timing Jitter in Inverters Induced by Power-Supply Noise," *International Conference on Design and Test of Integrated Systems in Nanoscale Technology*, pp. 53–56, 2006.

[37] J. A. McNeill, "Jitter in Ring Oscillators," *IEEE Journal of Solid State Circuits*, vol. 32, no. 6, pp. 870–879, Jun. 1997.

[38] R. Navid, T. H. Lee, and R. W. Dutton, "Minimum Achievable Phase Noise of RC Oscillators," *IEEE Journal of Solid-State Circuits*, vol. 40, no. 3, pp. 630–637, Mar. 2005.

[39] A. Abidi and R. G. Meyer, "Noise in Relaxation Oscillators," *IEEE Journal of Solid State Circuits*, pp. 794–802, Dec. 1983.

[40] B. Razavi, "A Study of Phase Noise in CMOS Oscillators," *IEEE Journal of Solid State Circuits*, pp. 331–343, Mar. 1996.

[41] J. R. Westra, "High-Performance Oscillators and Oscillator Systems," PhD dissertation, Delft University of Technology, Delft University Press, 1998.

[42] S. L. J. Gierkink, "Control Linearity and Jitter of Relaxation Oscillators," PhD dissertation, University of Twente, 1999.

[43] D. B. Leeson, "A Simple Model of Feedback Oscillator Noise Spectrum," *Proceedings of the IEEE*, vol. 54, pp. 329–330, Feb. 1966.

[44] J. Craninckx, "Low-Noise Voltage-Controlled Oscillators Using Enhanced LC-Tanks," *IEEE Transactions on Circuits and Systems II: Express Briefs*, vol. 42, pp. 794–804, Dec. 1995.

[45] A. Hajimiri and T. H. Lee, "Design Issues in CMOS Differential LC Oscillators," *IEEE Journal of Solid-State Circuits*, vol. 34, no. 5, pp. 717–724, May 1999.

[46] D. Ham and A. Hajimiri, "Concepts and Methods in Optimization of Integrated LC VCOs," *IEEE Journal of Solid-State Circuits*, vol. 36, no. 6, pp. 896–909, Jun. 2001.

[47] J. Vig, "Quartz Crystal Resonators and Oscillators," IEEE Ultrasonics, Ferroelectrics and Frequency Control Society, Tech. Rep., 2004. [Online]. Available: www.ieee-uffc.org/frequency-control/learning-vig-tut.asp

[48] E. Vittoz, M. Degrauwe, and S. Bitz, "High Performance Crystal Oscillator Circuits: Theory and Application," *IEEE Journal of Solid-State Circuits*, vol. 23, no. 3, pp. 774–783, Jun. 1988.

[49] A. Hajimiri and T. H. Lee, "A General Theory of Phase Noise in Electrical Oscillators," *IEEE Journal of Solid-State Circuits*, vol. 33, no. 2, pp. 179–194, Feb. 1998.

[50] ——, "Corrections to 'A General Theory of Phase Noise in Electrical Oscillators'," *IEEE Journal of Solid-State Circuits*, vol. 33, p. 928, Jun. 1998.

[51] A. Hajimiri, S. Limotyrakis, and T. H. Lee, "Jitter and Phase Noise in Ring Oscillators," *IEEE Journal of Solid-State Circuits*, vol. 34, no. 6, pp. 790–804, Jun. 1999.

[52] L. Lu, Z. Tang, P. Andreani, A. Mazzanti, and A. Hajimiri, "Comments on 'Comments on "A General Theory of Phase Noise in Electrical Oscillators" '," *IEEE Journal of Solid-State Circuits*, vol. 43, no. 9, p. 2170, Sep. 2008.

[53] A. Mazzanti and P. Andreani, "Class-C Harmonic CMOS VCOs, with a General Result on Phase Noise," *IEEE Journal of Solid-State Circuits*, vol. 43, no. 12, pp. 2716–2728, Dec. 2008.

[54] J. Bank, "A Harmonic-Oscillator Design Methodology Based on Describing Functions," PhD dissertation, Chalmers University, Sweden, 2006.

[55] D. Murphy, J. J. Rael, and A. A. Abidi, "Phase Noise in LC Oscillators: A Phasor-Based Analysis of a General Result and of Loaded Q," *IEEE Journal of Solid-State Circuits*, vol. 57, no. 6, pp. 1187–1203, Jun. 2010.

[56] C. Samori, A. L. Lacaita, F. Villa, and F. Zappa, "Spectrum Folding and Phase Noise in LC Tuned Oscillators," *IEEE Transactions on Circuits and Systems II: Express Briefs*, vol. 45, no. 7, pp. 781–790, Jul. 1998.

[57] P. Andreani and A. Fard, "More on the $1/f^2$ Phase Noise Performance of CMOS Differential-Pair LC-Tank Oscillators," *IEEE Journal of Solid-State Circuits*, vol. 41, no. 12, pp. 2703–2712, Dec. 2006.

[58] P. Andreani, X. Wang, L. Vandi, and A. Fard, "A Study of Phase Noise in Colpitts and LC-Tank CMOS Oscillators," *IEEE Journal of Solid-State Circuits*, vol. 40, no. 5, pp. 1107–1118, May 2005.

[59] E. Hegazi, H. Sjland, and A. A. Abidi, "A Filtering Technique to Lower LC Oscillator Phase Noise," *IEEE Journal of Solid-State Circuits*, vol. 36, no. 12, pp. 1921–1930, Dec. 2001.

[60] M. Garampazzi, P. M. Mendes, N. Codega, D. Manstretta, and R. Castello, "Analysis and Design of a 195.6 dBc/Hz Peak FoM P-N Class-B Oscillator with Transformer-Based Tail Filtering," *IEEE Journal of Solid-State Circuits*, vol. 50, no. 7, pp. 1657–1668, Jul. 2015.

[61] L. Fanori and P. Andreani, "Class-D CMOS Oscillators," *IEEE Journal of Solid-State Circuits*, vol. 48, no. 12, pp. 3105–3119, 2013.

[62] D. Murphy, H. Darabi, and H. Wu, "A VCO with Implicit Common-Mode Resonance," *IEEE International Solid-State Circuits Conference (ISSCC)*, pp. 442–443, Feb. 2015.

[63] D. Murphy and H. Darabi, "A Complementary VCO for IoE That Achieves a 195dBc/Hz FOM and Flicker Noise Corner of 200kHz," *IEEE International Solid-State Circuits Conference (ISSCC)*, pp. 44–45, Feb. 2016.

[64] P. F. J. Geraedts, E. van Tuijl, E. A. M. Klumperink, G. J. M. Wienk, and B. Nauta, "A 90 uw 12MHz Relaxation Oscillator with a −162db FOM," *IEEE International Solid-State Circuits Conference (ISSCC)*, pp. 348–618, Feb. 2008.

[65] X. Yang, K. Xu, W. Wang, Y. Uchida, and T. Yoshimasu, "2-GHz Band Ultra-Low-Voltage lc-vco IC in 130nm CMOS Technology," *IEEE International Conference of Electron Devices and Solid–State Circuits*, pp. 1–2, Jun. 2013.

[66] P. Dubey, D. Belot, and S. Chatterjee, "Heterogeneous Coupled Ring Oscillator Arrays for Reduced Phase Noise at Lower Power Consumption," *IEEE Asian Solid State Circuits Conference (A-SSCC)*, pp. 365–368, Nov. 2012.

[67] C.-H. Park and B. Kim, "A Low-Noise, 900MHz VCO in 0.6-μm CMOS," *IEEE Journal of Solid-State Circuits*, vol. 34, no. 5, pp. 586–591, May 1999.

[68] M. M. Abdul-Latif and E. Sanchez-Sinencio, "Low Phase Noise Wide Tuning Range n-Push Cyclic-Coupled Ring Oscillators," *IEEE Journal of Solid-State Circuits*, vol. 47, no. 6, pp. 1278–1294, Jun. 2012.

[69] C. Li and J. Lin, "A 1 to 9 GHz Linear-Wide-Tuning-Range Quadrature Ring Oscillator in 130 nm CMOS for Non-Contact Vital Sign Radar Application," *IEEE Microwave and Wireless Components Letters*, vol. 20, no. 1, pp. 34–36, Jan. 2010.

[70] S. W. Park and E. Sanchez-Sinencio, "RF Oscillator Based on a Passive RC Bandpass Filter," *IEEE Journal of Solid-State Circuits*, vol. 44, no. 11, pp. 3092–3101, Nov. 2009.

[71] S. Levantino, L. Romano, S. Pellerano, C. Samori, and A. Lacaita, "Phase Noise in Digital Frequency Dividers," *IEEE Journal of Solid-State Circuits*, vol. 39, no. 5, pp. 775–784, May 2004.

[72] A. V. Oppenheim and R. W. Schafer, *Discrete-Time Signal Processing*, 2nd ed. Prentice-Hall, 1999.

[73] D. J. Kinniment and J. V. Woods, "Synchronization and Arbitration Circuits in Digital Systems," *Proceedings of the IEEE*, vol. 123, no. 10, pp. 961–966, Oct. 1976.

[74] D. G. Messerschmitt, "Synchronization in Digital System Design," *IEEE Journal on Selected Areas in Communications*, vol. 8, no. 8, pp. 1404–1419, Oct. 1990.

[75] D. J. Kinniment, "Synchronization and Arbitration in GALS," *Electronic Notes in Theoretical Computer Science*, vol. 245, pp. 85–101, 2009.

[76] S. Beer, R. Ginosar, R. Dobkin, and Y. Weizman, "MTBF Estimation in Coherent Clock Domains," *IEEE 19th International Symposium on Asynchronous Circuits and Systems*, 2013.

[77] N. Da Dalt, M. Harteneck, C. Sandner, and A. Wiesbauer, "On the Jitter Requirements of the Sampling Clock for Analog-to-Digital Converters," *IEEE Transactions on Circuits and Systems I: Regular Papers*, vol. 49, no. 9, pp. 1354–1360, Sep. 2002.

[78] A. Varzaghani, A. Kasapi, D. N. Loizos, S. Paik, S. Verma, S. Zogopoulos, and S. Sidiropoulos, "A 10.3-GS/s, 6-Bit Flash ADC for 10G Ethernet Applications," *IEEE Journal of Solid-State Circuits*, vol. 48, no. 12, pp. 3038–3048, Dec. 2013.

[79] L. Kull et al., "A 90GS/s 8b 667mW 64x Interleaved SAR ADC in 32nm Digital SOI CMOS," *IEEE International Solid-State Circuits Conference (ISSCC)*, pp. 378–379, Feb. 2014.

[80] M. El-Chammas and B. Murmann, "General Analysis on the Impact of Phase-Skew in Time-Interleaved ADCs," *IEEE Transactions on Circuits and Systems I: Regular Papers*, vol. 56, no. 5, pp. 902–910, May 2009.

[81] ——, "A 12-GS/s 81-mW 5-bit Time-Interleaved Flash ADC with Background Timing Skew Calibration," *IEEE Journal of Solid-State Circuits*, vol. 46, no. 4, pp. 838–847, Apr. 2011.

[82] B. Razavi, "Design Considerations for Interleaved ADCs," *IEEE Journal of Solid-State Circuits*, vol. 48, no. 8, pp. 1806–1817, Aug. 2013.

[83] N. L. Dortz, J. P. Blanc, T. Simon, S. Verhaeren, E. Rouat, P. Urard, S. L. Tual, D. Goguet, C. Lelandais-Perrault, and P. Benabes, "A 1.62GS/s Time-Interleaved SAR ADC with Digital Background Mismatch Calibration Achieving Interleaving Spurs Below 70dBfs," *IEEE International Solid-State Circuits Conference (ISSCC)*, pp. 386–388, Feb. 2014.

[84] R. Schreier and G. Temes, *Understanding Delta–Sigma Data Converters*. John Wiley & Sons, 2005.

[85] T. H. Lee and J. F. Bulzacchelli, "A 155MHz Clock Recovery Delay- and Phase-Locked Loop," *IEEE Journal of Solid-State Circuits*, vol. 27, no. 12, pp. 1736–1746, Dec. 1992.

[86] M. J. Park and J. Kim, "Pseudo-Linear Analysis of Bang-Bang-Controlled Timing Circuits," *IEEE Transactions on Circuits and Systems I: Regular Papers*, vol. 60, no. 6, pp. 1381–1394, Jun. 2013.

[87] N. Da Dalt, "Markov Chains-Based Derivation of the Phase Detector Gain in Bang-Bang PLLs," *IEEE Transactions on Circuits and Systems II: Express Briefs*, vol. 53, no. 11, pp. 1195–1199, Nov. 2006.

[88] G. Balamurugan and N. Shanbhag, "Modeling and Mitigation of Jitter in High-Speed Source-Synchronous Interchip Communication Systems," *The Thirty-Seventh Asilomar Conference on Signals, Systems Computers, 2003*, vol. 2, pp. 1681–1687, Nov. 2003.

[89] F. Rao and S. Hindi, "Frequency Domain Analysis of Jitter Amplification in Clock Channels," *2012 IEEE 21st Conference on Electrical Performance of Electronic Packaging and Systems*, pp. 51–54, Oct. 2012.

[90] A. Ragab, Y. Liu, K. Hu, P. Chiang, and S. Palermo, "Receiver Jitter Tracking Characteristics in High-Speed Source Synchronous Links," *Journal of Electrical and Computer Engineering*, vol. 2011, pp. 1–15, 2011.

[91] B. Casper and F. O'Mahony, "Clocking Analysis, Implementation and Measurement Techniques for High-Speed Data Links: A Tutorial," *IEEE Transactions on Circuits and Systems I: Regular Papers*, vol. 56, no. 1, pp. 17–39, Jan. 2009.

[92] T. Takemoto, H. Yamashita, F. Yuki, N. Masuda, H. Toyoda, N. Chujo, Y. Lee, S. Tsuji, and S. Nishimura, "A 25-Gb/s 2.2-W 65-nm CMOS Optical Transceiver Using a Power-Supply-Variation-Tolerant Analog Front End and Data-Format Conversion," *IEEE Journal of Solid-State Circuits*, vol. 49, no. 2, pp. 471–485, Feb. 2014.

[93] H. Noguchi, N. Yoshida, H. Uchida, M. Ozaki, S. Kanemitsu, and S. Wada, "A 40-Gb/s CDR Circuit with Adaptive Decision-Point Control Based on Eye-Opening Monitor Feedback," *IEEE Journal of Solid-State Circuits*, vol. 43, no. 12, pp. 2929–2938, Dec. 2008.

[94] K. Nose, M. Kajita, and M. Mizuno, "A 1-ps Resolution Jitter-Measurement Macro Using Interpolated Jitter Oversampling," *IEEE Journal of Solid-State Circuits*, vol. 41, no. 12, pp. 2911–2920, Dec. 2006.

[95] T. Hashimoto, H. Yamazaki, A. Muramatsu, T. Sato, and A. Inoue, "Time-to-Digital Converter with Vernier Delay Mismatch Compensation for High Resolution On-Die Clock Jitter Measurement," *IEEE Symposium on VLSI Circuits*, pp. 166–167, Jun. 2008.

[96] J. Liang, A. Sheikholeslami, H. Tamura, Y. Ogata, and H. Yamaguchi, "A 28Gb/s Digital CDR with Adaptive Loop Gain for Optimum Jitter Tolerance," *IEEE International Solid-State Circuits Conference (ISSCC)*, pp. 122–123, Feb. 2017.

[97] J. Liang, M. S. Jalali, A. Sheikholeslami, M. Kibune, and H. Tamura, "On-Chip Measurement of Clock and Data Jitter with Sub-Picosecond Accuracy for 10 Gb/s Multilane CDRs," *IEEE Journal of Solid-State Circuits*, vol. 50, no. 4, pp. 845–855, Apr. 2015.

[98] H. Takauchi, H. Tamura, S. Matsubara, M. Kibune, Y. Doi, T. Chiba, H. Anbutsu, H. Yamaguchi, T. Mori, M. Takatsu, K. Gotoh, T. Sakai, and T. Yamamura, "A CMOS Multichannel 10Gb/s Transceiver," *IEEE Journal of Solid-State Circuits*, vol. 38, no. 12, pp. 2094–2100, Dec. 2003.

[99] J. Liang, A. Sheikholeslami, H. Tamura, and H. Yamaguchi, "Jitter Injection for On-Chip Jitter Measurement in PI-Based CDRs," *IEEE Custom Integrated Circuits Conference (CICC)*, pp. 1–4, Apr. 2017.

[100] J. T. J. Penttinen, Ed., *The Telecommunications Handbook: Engineering Guidelines for Fixed, Mobile and Satellite Systems*. John Wiley & Sons, 2015.

[101] A. Liscidini, L. Fanori, P. Andreani, and R. Castello, "A Power-Scalable DCO for Multi-Standard GSM/WCDMA Frequency Synthesizers," *IEEE Journal of Solid-State Circuits*, vol. 49, no. 3, pp. 646–656, Mar. 2014.

[102] M. Mikhemar, D. Murphy, A. Mirzaei, and H. Darabi, "A Phase-Noise and Spur Filtering Technique Using Reciprocal-Mixing Cancellation," *IEEE International Solid-State Circuits Conference (ISSCC)*, pp. 86–87, Feb. 2013.

[103] H. M. Walker and J. Lev, *Statistical Inference*. Holt Rinehart & Winston, 1953.

[104] S. S. Wilks, *Mathematical Statistics*. John Wiley & Sons, 1962.

[105] W. W. Hines, D. C. Montgomery, D. M. Goldsman, and C. M. Borror, *Probability and Statistics in Engineering*. John Wiley & Sons, 2003.

[106] S. S. Wilks, "Determination of Sample Sizes for Setting Tolerance Limits," *The Annals of Mathematical Statistics*, vol. 12, pp. 91–96, 1941.

[107] H. Freeman, *Introduction to Statistical Inference*. Addison-Wesley, 1962.

[108] J. Rutman and F. L. Walls, "Characterization of Frequency Stability in Precision Frequency Sources," *Proceedings of the IEEE*, vol. 79, no. 6, pp. 952–960, Jun. 1991.

[109] D. Sullivan, D. Allan, D. Howe, and F. Walls, "Characterization of Clocks and Oscillators," National Institute of Standards and Technology (NIST), Tech. Rep., Jan. 1990, Technical Note 1337.

[110] W. J. Riley, "Handbook of Frequency Stability Analysis," National Institute of Standards and Technology (NIST), Tech. Rep., Jul. 2008, Special Publication 1065.

[111] I. Flinn, "Extent of the 1/f Noise Spectrum," *Nature*, vol. 219, pp. 1356–1357, Sep. 1968.

[112] Y. Tsividis, *Operation and Modeling of the MOS Transistor*. Oxford University Press, 1999.

[113] M. A. Caloyannides, "Microcycle Spectral Estimates of 1/f Noise in Semiconductors," *Journal of Applied Physics*, vol. 45, pp. 307–316, 1974.

[114] W. Schottky, "Small-Shot Effect and Flicker Effect," *Physics Review*, vol. 28, pp. 74–103, 1926.

[115] J. Bernamont, "Fluctuations in the Resistance of Thin Films," *Proceedings of the Physical Society*, vol. 49, pp. 138–139, 1937.

[116] B. B. Mandelbrot, "Some Noises with l/f Spectrum, a Bridge Between Direct Current and White Noise," *IEEE Transactions on Information Theory*, vol. 13, no. 2, pp. 289–298, 1967.

[117] J. J. Brophy, "Low-Frequency Variance Noise," *Journal of Applied Physics*, vol. 41, pp. 2913–2916, 1970.

[118] M. S. Keshner, "1/f Noise," *Proceedings of the IEEE*, vol. 70, no. 3, pp. 212–218, Mar. 1982.

[119] I. S. Gradshteyn and I. M. Ryzhik, *Table of Integrals, Series, and Products*. Academic Press, 2007.

[120] V. Radeka, "1/f Noise in Physical Measurements," *IEEE Transactions on Nuclear Science*, pp. 17–35, Nov. 1969.

[121] D. Ham and A. Hajimiri, "Virtual Damping and Einstein Relation in Oscillators," *IEEE Journal of Solid-State Circuits*, vol. 38, no. 3, pp. 407–418, Mar. 2003.

[122] M. S. Keshner, "Renewal Process and Diffusion Models of 1/f Noise," PhD dissertation, Massachusetts Institute of Technology, 1975.

[123] F. X. Kaertner, "Analysis of White and $f^{-\alpha}$ Noise in Oscillators," *International Journal of Circuit Theory and Applications*, vol. 18, pp. 485–519, 1990.

[124] D. C. Montgomery and E. A. Peck, *Introduction to Linear Regression Analysis*. John Wiley and Sons, 1991.

[125] A. C. Popovici, "Fast Measurement of Bit Error Rate in Digital Links," *IEE Proceedings*, vol. 134, no. 5, pp. 439–447, Aug. 1987.

[126] M. Miller, "Estimating Total Jitter Concerning Precision, Accuracy and Robustness," *DesignCon*, Feb. 2007.

[127] ——, "Measuring Components of Jitter," *US Patent 7 516 030*, Apr. 2009.

[128] S. Erb and W. Pribyl, "Design Specification for BER Analysis Methods Using Built-In Jitter Measurements," *IEEE Transactions on Very Large Scale Integration (VLSI) Systems*, vol. 20, no. 10, Oct. 2012.

[129] W. A. Gardner, *Introduction to Random Processes with Applications to Signals and Systems*, 2nd ed. McGraw-Hill, 1989.

[130] D. Halford, J. H. Shoaf, and A. S. Risley, "Spectral Density Analysis: Frequency Domain Specification and Measurement of Signal Stability," *27th Annual Symposium on Frequency Control*, pp. 421–431, Jun. 1973.

Index

Printed in the United States
By Bookmasters